Springer Theses

Recognizing Outstanding Ph.D. Research

Aims and Scope

The series "Springer Theses" brings together a selection of the very best Ph.D. theses from around the world and across the physical sciences. Nominated and endorsed by two recognized specialists, each published volume has been selected for its scientific excellence and the high impact of its contents for the pertinent field of research. For greater accessibility to non-specialists, the published versions include an extended introduction, as well as a foreword by the student's supervisor explaining the special relevance of the work for the field. As a whole, the series will provide a valuable resource both for newcomers to the research fields described, and for other scientists seeking detailed background information on special questions. Finally, it provides an accredited documentation of the valuable contributions made by today's younger generation of scientists.

Theses are accepted into the series by invited nomination only and must fulfill all of the following criteria

- They must be written in good English.
- The topic should fall within the confines of Chemistry, Physics, Earth Sciences, Engineering and related interdisciplinary fields such as Materials, Nanoscience, Chemical Engineering, Complex Systems and Biophysics.
- The work reported in the thesis must represent a significant scientific advance.
- If the thesis includes previously published material, permission to reproduce this must be gained from the respective copyright holder.
- They must have been examined and passed during the 12 months prior to nomination.
- Each thesis should include a foreword by the supervisor outlining the significance of its content.
- The theses should have a clearly defined structure including an introduction accessible to scientists not expert in that particular field.

More information about this series at http://www.springer.com/series/8790

Jan Philipp Dabruck

Target Station Optimization for the High-Brilliance Neutron Source HBS

Simulation Studies Based on the Monte Carlo Method

Doctoral Thesis accepted by
the RWTH Aachen University, Aachen,
Germany

Springer

Author
Dr. rer. nat. Jan Philipp Dabruck
Faculty of Georesources and Materials Engineering, Institute for Nuclear Engineering and Technology Transfer (NET)
RWTH Aachen University
Aachen, Germany

Supervisors
Prof. Dr. Bruno Thomauske
Faculty of Georesources and Materials Engineering, Institute for Nuclear Engineering and Technology Transfer (NET)
RWTH Aachen University
Aachen, Germany

Prof. Dr. rer. nat. Rahim Nabbi
Faculty of Georesources and Materials Engineering, Institute for Nuclear Engineering and Technology Transfer (NET)
RWTH Aachen University
Aachen, Germany

Prof. Dr. rer. nat. Achim Stahl
Faculty of Mathematics, Computer Science and Natural Sciences, Physics Institute III B
RWTH Aachen University
Aachen, Germany

ISSN 2190-5053 ISSN 2190-5061 (electronic)
Springer Theses
ISBN 978-3-030-05638-4 ISBN 978-3-030-05639-1 (eBook)
https://doi.org/10.1007/978-3-030-05639-1

Library of Congress Control Number: 2018964022

This Springer imprint is published by the registered company Springer Nature Switzerland AG
The registered company address is: Gewerbestrasse 11, 6330 Cham, Switzerland

Study hard what interests you the most in the most undisciplined, irreverent, and original manner possible.

Richard Feynman

Supervisor's Foreword

Neutron scattering has become a powerful and popular tool in many areas of science and industry, especially where X-radiation and other non-invasive methods cannot provide sufficient results. It is therefore to be expected that the demand for beam time at neutron sources will continue to increase. This contrasts with the fact that more and more research reactors in Europe are shut down, which are the most productive and reliable source of intense neutron radiation. With the European Spallation Source (ESS), an accelerator-driven spallation neutron source currently being under construction, it is intended to try to solve this problem. However, it is obvious that the entire European demand for beam time cannot be covered by few large high-power sources.

Therefore, a new concept had been developed that a network of medium and small sources should be established in Europe to complement the large neutron sources. To this end, Forschungszentrum Jülich is pushing ahead with the development of a compact accelerator-driven medium-flux neutron source in the sub-spallation regime called "High-Brilliance Neutron Source" (HBS). The central goal is to maximize the neutron yield of such sub-spallation sources by conducting intense optimization studies on all components.

The moderator is the key element of a highly brilliant neutron source. The development and optimization of the moderator system is the central task of Mr. Dabruck's Ph.D. thesis. The results should be considered primarily in the development of HBS, but may also be used in the design of other sources of the targeted network. In a first step, Mr. Dabruck has implemented a realistic target geometry and investigated the reasonable energy range of the ion accelerator, approximately 5–100 MeV, with regard to the neutron yield to be achieved.

For two possible configurations, 25 MeV deuterons and 50 MeV protons, Mr. Dabruck examined all possible geometric arrangements and dimensions for typical moderator materials. The focus was on maximizing the thermal neutron flux inside the moderator. From the results, he derived a moderator configuration, which consists of a small-volume hydrogen-based moderator and a large-volume reflector made of beryllium. Afterwards, Mr. Dabruck modified this arrangement with the aim of maximizing the brilliance, i.e., the neutron flux per energy range and solid angle available to the instruments. In cooperation with the project partners, he developed and analyzed the concept of "flux channels," by which the thermal neutrons can be extracted directly from the center of the moderator system. This significantly increases the brilliance for the downstream instruments. This concept led to a European patent due to its efficiency and novel character. Mr. Dabruck also equipped the moderator system with several cold moderators operated at very low temperatures around 20 K, which have been optimized for different materials, such as para-hydrogen or aromatic substances, in their arrangement and their dimensions.

Another subject of Mr. Dabruck's research was the optimization of a moderator system for a neutron source based on the laser-induced acceleration of ions. Such acceleration mechanisms are subject of current research and will play a major role in the future, as they allow for much more compact devices. In addition, Mr. Dabruck also carried out optimization studies for a prototype moderator that was able to demonstrate the efficiency of the flux channels at the training reactor AKR-2 in Dresden.

The analyses of Mr. Dabruck described in his thesis represent a great leap forward in the development of High-Brilliance Neutron Sources.

Aachen, Germany
October 2018

Prof. i.R. Bruno Thomauske

Abstract

The importance of neutrons as scientific probes is undisputed. The HBS, planned by Forschungszentrum Jülich, shall serve as a precursor for a network of low- and medium-flux neutron sources throughout Europe and will make use of the neutron-producing reaction Be(p,n) or Be(d,n) in the sub-spallation regime with $E_{\mathrm{ion}} \lesssim 100$ MeV.

Subject of this thesis is the development and optimization of the target-moderator-reflector system of the HBS. The dependence of the brilliance and temporal pulse structure of the generated neutrons on the geometry, the materials, and the accelerator parameters is examined in multi-dimensional parameter studies. The investigations are conducted by means of the Monte Carlo method.

The simulation studies indicate that a combination of a compact thermal moderator with a high slowing-down power $\xi\Sigma_{\mathrm{el}}$, for example, PE or light water, and a high-volume reflector with a low capture cross section $\sigma_{(\mathrm{n},\gamma)} \ll \sigma^{\mathrm{H}}_{(\mathrm{n},\gamma)}$, e.g., beryllium or graphite, exhibits an exceptionally high thermal neutron flux within the moderator-reflector assembly. The neutrons can effectively be extracted by exploiting the streaming effect using newly developed flux channels from the center of the moderator, whereby the peak brilliance is increased significantly. Cold moderators should be implemented in the flux maximum inside the thermal moderator to additionally increase the peak brilliance of cold neutrons. For the mechanical engineering of the target-moderator-reflector assembly, a design of all components as compact as possible is strongly advised.

A further subject of this thesis is the design and optimization of a moderator prototype for measurements at the AKR-2 training reactor in Dresden, by which the conclusions of the simulations are verified.

Acknowledgements

At this point, I would like to express my thanks to my supervisors Prof. Thomauske and Prof. Nabbi. Thanks to his professional knowledge and his many years of experience, Prof. Nabbi was always able to give me new impulses and was always open and committed to technical discussions and problems. Professor Thomauske always had the big picture in mind and has therefore always been a help when my next steps were uncertain.

I would like to thank Frank Charlier for his support, his confidence, and for always ensuring a relaxed and pleasant working atmosphere. Only this allowed me to undertake creative and productive work.

I would also like to thank Prof. Brückel and the staff of the JCNS-2 for the excellent cooperation in this great project.

My office colleagues, I thank for very stimulating discussions and the one or other exhilarating distraction from everyday work.

My deepest thanks go to my parents Regine and Andreas and my sisters Rebecca and Luisa, who have always supported me and found the right words to let me continue even in difficult times.

Contents

Acronyms

AKR-2	Ausbildungskernreaktor-2
BNCT	Boron-neutron-capture therapy
BOA	Breakout Afterburner
CANS	Compact accelerator-driven neutron source
DPA	Displacements per atom
ENDF	Evaluated Nuclear Data File
ESS	European Spallation Source
FWHM	Full width at half maximum
HBS	High-Brilliance Neutron Source
iWASP	Ion wide angle spectrometer
JCANS	Japan Collaboration on Accelerator-driven Neutron Sources
MPI	Message Passing Interface
NAIS	Nuclear activation-based imaging spectroscopy
OMP	Open Multi-Processing
PKA	Primary Knock-on Atom
RANS	Riken Accelerator-Driven Neutron Source
RFQ	Radio frequency quadrupole
SANS	Small-angle neutron scattering
SDEF	Source definition
TENDL	TALYS Evaluated Nuclear Data Library
TNSA	Target Normal Sheath Acceleration
TOF	Time-of-flight

Chapter 1
Introduction

Neutrons are and will become increasingly important in the future with many applications in science, industry, nuclear medicine, and nuclear chemistry.

For applications such as scattering methods and radiography, neutrons offer some advantageous properties when compared to X-ray radiation. While X-ray radiation interacts with the electron cloud, neutrons interact with the atomic nuclei and are thus also sensitive to light isotopes, especially hydrogen, as well as with the magnetic moment of the sample atoms via their spin. Thermal neutrons can be used to analyze the crystalline structure of solids, e.g., in diffractometry and 3-axis spectrometry. Cold neutrons can be used to identify organic molecules by their vibrational and rotational degrees of freedom (spin echo spectroscopy, reflectometry, and diffractometry) or to investigate magnetic nanostructures by *small-angle neutron scattering* (SANS) [1, 2]. Since neutrons carry no electric charge they deeply penetrate into matter and can be used as a non-destructive probe for large samples [3]. Impressive examples are the radiography of a running engine block or the investigation of archaeological finds via neutron imaging [4]. Further important applications are the production of radionuclides [5, 6], which in turn have a wide range of uses, the neutron activation analysis [7], which is suitable for distinguishing individual isotopes due to the characteristic gamma radiation emitted when capturing thermal neutrons, as well as cancer treatment with fast neutrons or the *boron-neutron-capture therapy* (BNCT) with thermal neutrons [8].

As a drawback, the generation of free neutrons is very complex. The basic mechanisms are always nuclear reactions, either (spontaneous or induced) nuclear fission, radioactive decay or the interaction of stable nuclides with high-energy ions or electrons. The highest source strengths and neutron fluxes are achieved at research fission reactors with the reactor at the Institut Laue-Langevin (ILL) in Grenoble, France, being the most powerful neutron source providing 49 different instruments. According to the thermal power of 58.3 MW the neutron source strength amounts to 5×10^{18} n/s [9]. The strongest research reactor in Germany is the FRM 2 at the

J. P. Dabruck, *Target Station Optimization for the High-Brilliance Neutron Source HBS*, Springer Theses, https://doi.org/10.1007/978-3-030-05639-1_1

Heinz Maier-Leibnitz Zentrum in Munich with a thermal power of 20 MW offering a source strength of 1.7×10^{18} n/s and providing 39 experiments [10].

In 2011 the 13th amendment of the atomic energy act was passed which purports the phase-out of nuclear power in Germany. However, not only nuclear power plants are affected, but also German research reactors. After the upcoming decommissioning of the Berliner Experimentier-Reaktor 2 and TRIGA in Mainz, the FRM II remains the only powerful neutron source in Germany. Additionally, the inadequate availability of highly enriched uranium also means that more and more research reactors are being decommissioned throughout Europe, for example the 23 MW heavy water moderated FRJ–2 in Jülich, Germany, decommissioned in 2006. In the midterm a bottleneck in the supply with radionuclides for nuclear medicine, e.g., ^{99m}Tc, as well as beam time for neutron scattering experiments is expected.

The only way to substitute high-performance research reactors are spallation sources. Spallation is a process which occurs when high-Z materials are bombarded with high-energy protons ($E \gtrsim 100$ MeV) whereby the target nuclei are highly excited and emit up to 30 neutrons in a two-stage process. The European Spallation Source ESS, which is under construction in Lund, Sweden, is expected to be operational in 2019 and will be the world's strongest spallation source with a source strength of 10^{18}n/s [11].

Nonetheless, high-flux sources are and will always be overbooked with overloading factors of typically 2–3 and can supply only few experiments. The recruitment and education of junior scientists as well as instrument development is therefore almost impossible at such facilities. Low-flux sources such as radionuclide sources and commercial fusion-based neutron generators are suitable for some applications, but inapplicable for most types of scattering experiments and radionuclide production due to the insufficient neutron fluxes [12].

To close the supply gap of beamtime at medium-flux sources, a network of *compact accelerator-driven neutron source* (CANS) shall be established in Europe similar to the *Japan Collaboration on Accelerator-driven Neutron Sources* (JCANS). For this aim, the project *High-Brilliance Neutron Source* (HBS) was commenced by the Jülich Center for Neutron Science 2/peter Grünberg Institute 4, Forschungszentrum Jülich, Germany, which is intended to serve as an initiator for such a network and will operate in the sub-spallation regime ($E \lesssim 100$ MeV) based on a light ion accelerator and a light metal target, for example, by exploiting the nuclear reactions Be(p,n) or Be(d,n). These reactions are less efficient than spallation regarding the neutron yield, being in the range of 1% per initial ion depending on the ion energy. To counter this disadvantage the entire facility is optimized in terms of neutron brilliance, which is a measure of how many neutrons are actually available to the experiments. Compared to spallation sources, there are several advantages that partially compensate for the lower efficiency. The considerably lower accelerator power not only reduces costs for acquisition, installation, and operation, but the layout of the target and thus also of the moderator is significantly more compact, which benefits the brilliance. Due to the markedly lower dose rate, the instrumentation can be installed much closer to the moderator, which increases the amount of neutrons that can be used for experiments. Moreover, the reduction in the need for shielding leads to further cost reduction.

Because of the good scalability between 1 and 100 MeV CANS are suitable not only for larger research facilities but also for, e.g., universities.

Since CANS still represent a comparatively young type of neutron sources, the associated moderators contain a large potential for optimization. However, the experiences gained from spallation sources with respect to the performance of the moderator cannot be transferred to CANS because of their much higher energy and power, and the resulting larger dose rates and higher cooling demands of the target. The completely different functional principle and the sheer size of research reactors also require a different moderator design, since its main task is the maintenance of the chain reaction.

The acceleration of light ions can also be accomplished with high-intensity short-pulse lasers which was favored at the beginning of the project HBS, since such lasers allow for an even more compact design of the facility. However, even though the development of such lasers has made tremendous progress in recent years, it has become clear that until the planned commissioning of the HBS it is not ensured that the reliability of such sources is sufficient for a regular operation since the optical components are not yet able to withstand the high continuous loads. Thus, in the course of the project the decision was made to rely on conventional light-ion accelerators.

In this thesis, simulation studies are presented which aim at optimizing the core components crucial for the neutronic performance of a laser- or accelerator-driven neutron source. The neutron flux in the moderator and the peak brilliance at the moderator surface is maximized taking into account the scope conditions dictated by the instruments to be operated at the facility. These performance quantities are hence determined at the entrance of the instrumentation for each neutron beamline. The simulations are carried out using the well-established Monte Carlo code MCNP, which is capable of adequately simulating the microscopic interactions of particles, including neutrons, photons, protons, and electrons, with matter.

In the Chap. 2 of this thesis the relevant physical quantities are discussed as well as the physical and technical principles of the processes in the components in question, beginning with the accelerator and the laser, respectively, the target, the moderator-reflector assembly, up to the neutron instrumentation. Concluding this chapter, the fundamentals of the simulation methods applied for this work are explained. The Chap. 3 covers the optimization studies on a laser-driven system. At the extremely powerful Trident laser located at the Los Alamos National Laboratory, USA, various promising experiments were conducted with the aim to produce neutrons. Thus, the framework parameters of the simulations discussed in this chapter are based on these experimental results. The Chap. 4 elucidates the development of a target-moderator-reflector assembly, which is referred to as the target station taking also the shielding and the supporting structures into account, for a CANS facility. Some results and conclusions from the previous work on the laser-driven facility are transferred, so that the optimum adaptation of the neutron source to the demands of the instruments is focused on. The simulation studies are verified at the AKR-2 (Ausbildungskernreaktor-2) training reactor in Dresden, Germany, by means of a heavy water prototype moderator containing a cryogenic moderator. For this purpose simulation studies are performed to optimize the prototype assembly for usage at this low-flux reactor, as presented in the Chap. 5.

The investigations accomplished show that the neutron output and brilliance of the thermal and cold beamlines can be increased significantly by optimizing the target-moderator-reflector assembly.

References

1. C.G. Shull, Early development of neutron scattering. Rev. Mod. Phys. **67**(4), 753–757 (1995). https://doi.org/10.1103/RevModPhys.67.753
2. A. Furrer, J. Mesot, T. Strässle, *Neutron Scattering in Condensed Matter Physics*, vol. 4, 1st edn. Neutron Techniques and Applications (World Scientific, 2009). ISBN: 978-9810248314
3. I.S. Anderson, R.L. McGreevy, H.Z. Bilheux, *Neutron Imaging and Applications: A Reference for the Imaging Community* (Springer Science, New York, 2009). ISBN 978-0-387-78692-6. https://doi.org/10.1007/978-0-387-78693-3
4. A.M. Pollard, C. Heron, *Archaeological Chemistry* (Royal Society of Chemistry, Cambridge, 1996). ISBN 0-85404-523-6
5. K.H. Beckurts, K. Wirtz, *Neutron Physics* (Springer, Heidelberg, 1964), pp. 28–29
6. P.A. Balo, J.B. Knauer, R.C. Martin, Production, distribution, and applications of californium-252 neutron sources, in *IRRMA'99: 4th Topical Meeting on Industrial Radiation*, Raleigh, NC (US), 3–7 Oct 1999 (ORNL/CP-102606). (Oak Ridge National Lab, TN (US) 1999). http://www.osti.gov/scitech/servlets/purl/15053
7. P. Kruger, *Principles of Activation Analysis*. (Wiley, New York, 1971). ISBN: 978-0471508601
8. R.F. Barth, A.H. Soloway, R.G. Fairchild, Boron neutron capture therapy for cancer. Sci. Am. **263**(4), 100–107 (1990). https://doi.org/10.1038/scientificamerican1090-100
9. G. Cicognani, H. Schober. ILL Annual Report. (Institut Laue- Langevin, Grenoble, France, 2015). https://www.ill.eu/fileadmin/users_files/Annual_Report/AR-15/mobile/index.html#p=1. Accessed 27 June 2017
10. A. Kastenmüller, C. Kortenbruck, A. Voit, 10 Jahre Forschungs- Neutronenquelle Heinz Maier-Leibnitz (FRM II) (Technische Universitänchen, Deutschland, 2014). http://www.frm2.tum.de/fileadmin/w00bnv/www/Aktuelles___Medien/Broschueren/Sonstige/20140328_Festschrift_komplett_online_.pdf. Accessed 27 June 2017
11. S. Peggs et al., ESS Technical Design Report. ESS-0016915. European Spallation Source, Lund, Sweden, 23 Apr 2013. http://europeanspallationsource.se/documentation/tdr.pdf. Accessed 27 June 2017
12. T. Brückel, D. Richter, G. Roth, Andreas Wischnewski, R. Zorn (eds.), *Neutron Scattering: Lectures*, vol. 106. Schlüsseltechnologien/Key Technologies. (Forschungszentrum Jülich GmbH, 2015). ISBN: 978-3-95806-055-5

Chapter 2
Physical and Technical Principles

All processes occurring in nature are based on interactions between elementary particles mediated by the four fundamental forces (electromagnetic, weak, and strong interaction, and gravitation). For composite particles, processes occur which can be regarded as effective interactions. Examples are nuclear reactions of neutrons in matter, the Bremsstrahlung emission of charged particles in the Coulomb field of atomic nuclei, or the radioactive decay of unstable isotopes. The underlying processes are of a stochastic nature and the probability of their occurrence is parametrized by the *cross section* [1, 2].

The cross section is highly specific for each reaction channel of all possible particle-target combinations and also shows a characteristic dependence on the energy of the incident particle. For neutrons, the cross section can be divided into three regions. At low neutron energies, the cross section is inversely proportional to the neutron velocity, which is why this range is also called the $1/v$-range. In the case of medium energies, in the so-called resonance region, especially heavy target nuclei show strongly pronounced resonances. High energies represent the fast range in which the cross section gradually decreases and approaches double the value of the geometric cross-sectional area of the nucleus. For most other incident particles there is no strong regularity, so each reaction channel has a very characteristic shape.

In this chapter the functionality and mechanisms of the individual components that are relevant for CANS as well as the cross sections of the underlying processes are discussed.

Beginning with the acceleration of the ions, recent experiments using high-intensity lasers and suitable types of conventional particle accelerators are reviewed. The ion-target interaction is examined in terms of the neutron producing reactions and the target integrity which is affected by the bombardment with the high-energy ions.

The principles of neutron moderation are discussed concerning thermalization and the production of cold neutrons as well as the properties of common moderator materials. The fundamental quantity for assessing the performance of a thermal or

J. P. Dabruck, *Target Station Optimization for the High-Brilliance Neutron Source HBS*, Springer Theses, https://doi.org/10.1007/978-3-030-05639-1_2

cold moderator for most types of experiments is given by the brilliance. The brilliance measures how many neutrons of a given energy and direction are concentrated on a spot per unit area and is defined as $B = \frac{\Delta\Phi}{\Delta\Omega\,\Delta E}$. Typically, the brilliance is determined in a cone with half an apex angle θ with respect to the direction of the beamline, i.e., the forward direction. The corresponding solid angel is thus given as $\Omega = 2\,\pi\,(1 - \cos\theta)$.

Important components of the subsequent instrumentation of the experiments are also explained in this chapter, since their properties determine the parameters of the moderator system to be optimized.

2.1 Acceleration of Ions

For the investigations presented in this thesis, protons and deuterons are of particular interest. This section focuses on the mechanisms and principles for accelerating these types of particles.

2.1.1 Laser-Induced Acceleration

With the development of high-intensity ultra-short-pulsed lasers, a new efficient way of ion acceleration has emerged.

As individual photons of such lasers have energies in the range of 1 eV, it has to be taken into account that for sufficiently high intensities the interaction of light with matter is completely governed by the electromagnetic field rather than by single photons. Today's most powerful lasers are capable of accelerating ions up to hundreds of MeV [3] as described hereafter. An incident laser beam generates plasma on the target surface. Depending on the performance of the laser and the configuration of the experimental setup the ions and electrons of the plasma can be accelerated by three different mechanisms that lead to different results with regard to ion flux and spectrum.

Ponderomotive Acceleration In the electromagnetic field of the laser pulse, electrons and ions are affected by Lorentz-force. Due to the approximately Gaussian-like transverse spatial intensity profile, electrons travel along the intensity gradient which is referred to as ponderomotive force [3, 4]. The ponderomotive force leads to a charge separation with an extension of the size of the focal spot (typically 10 μm), so that a strong electric potential is established which isotropically accelerates the ions away from the focal spot.

Target Normal Sheath Acceleration For a target thickness of size of the focal spot (10 μm), some electrons extend past the rear surface of the target, forming an electrostatic sheath on this surface as shown in Fig. 2.1. The quasi-stationary ions are thus accelerated towards the electrons according to Coulomb's law [5–9]. It is

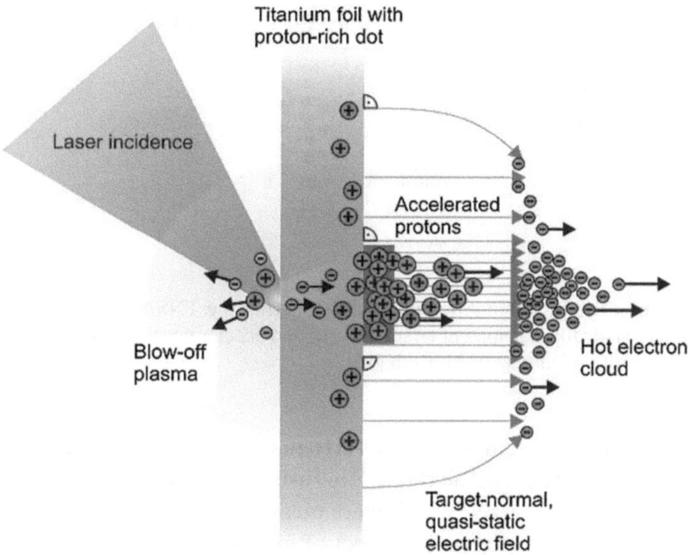

Fig. 2.1 Sketch of the Target Normal Sheath Acceleration (TNSA). For sufficiently thin targets electrons are driven out of the target by the laser forming an electrostatic sheath on the target surface which in turn accelerates the quasi-stationary ions [11]

important to note that the Target Normal Sheath Acceleration highly depend on the mass to charge ratio, consequently protons are preferred over all heavier ions such as deuterons [10].

Breakout Afterburner For high intensities and very thin targets, the target material becomes relativistically transparent for the laser light, so that the laser beam interacts with the whole illuminated target volume. This regime is called *Breakout Afterburner* (BOA) [12–14] and schematically depicted in Fig. 2.2. Due to the ponderomotive acceleration, electrons are driven through the target and a rear side sheath is generated. During the laser pulse, more and more electrons are heated up to relativistic energies so that in the case of a sufficiently thin target, the material becomes relativistically transparent for the laser beam. In this way, the laser can accelerate the rear-side electrons farther in beam direction, resulting in a much longer acceleration path for the ions. It should be noted, that, in contrast to TNSA, the BOA mechanism is independent of the charge to mass ratio [15].

Recent Experiments As an example of TNSA-based neutron production an experiment carried out by [10] at the ATLAS laser of the Max Planck Institute of Quantum Optics in Garching, Germany. New amplifier modules and adaptive optics were installed for the stage ATLAS-10 of the Ti:Sapphire laser. ATLAS-10 is currently capable of delivering 700 mJ, 160 fs pulses at a mean wavelength of 790 nm and 10 Hz repetition rate to the evacuated target chamber. The power therefore amounts to 4.3 TW which is focusable to an intensity of 2×10^{19} W/cm^2. The target was a CD_2

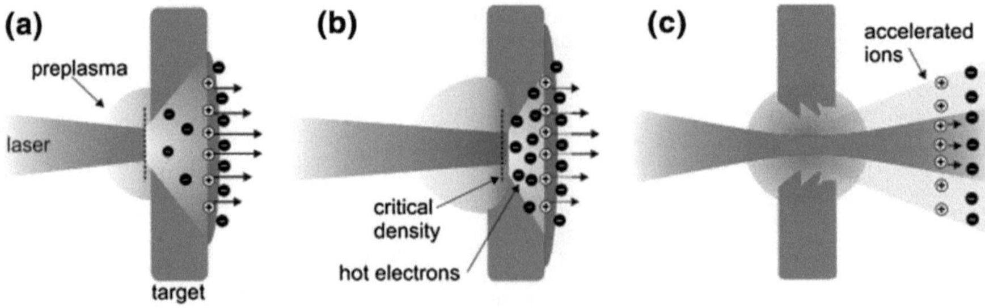

Fig. 2.2 Sketch of the Breakout Afterburner (BOA). After the TNSA phase (**a**) the laser breaks through the relativistically transparent target film (**b**) and accelerates the rear-site electrons further in the BOA phase (**c**). Courtesy of M. Roth [16]

disk with a diameter of 5 cm and a few 100 μm thickness. The lifetime of the target disks was about 6 000 shots. The $d(d,n)^3He$ fusion reaction produced 15 000–25 000 neutrons per shot in the target, which were measured by TOF detector arrays.

In order to further investigate the effect of *proton poisoning*, where the deuteron plasma is contaminated by protons, which are accelerated predominantly due to their smaller mass-to-charge ratio and shielding the field gradient, further experiments are carried out by [10] at the Friedrich Schiller University in Jena. The Ti:Sapphire laser can deliver pulses of 80 fs and 600 mJ of energy at a repetition rate of 10 Hz. The intensity amounts to $3 \times 10^{19}\,W/cm^2$ on a spot with a FWHM (full width at half maximum) diameter of 3 μm. The laser irradiated a D_2O jet which was produced with a nozzle size of 10 μm and a pressure of 50 bar. The jet was broken up into droplets of 20 μm diameter spaced by 80 μm by a piezo ceramic element operating at 1 MHz. Behind the D_2O droplet jet an additional CD_2 catcher was mounted in order to increase the neutron output by additional d-d fusion processes. Since the measured neutron spectra were directly used to determine the ion distribution, no absolute neutron sources strength was measured. It was found, however, that proton impurities have a high influence on the neutron production with deuterated targets, so that pure proton-beams can even be advantageous.

In order to investigate the scalability of the neutron yield as a function of the intensity of the laser used, the experimental approach was transferred to a significantly more powerful laser. At the 100 TW laser LULI (Laboratoire pour l'Utilisation des Laser Intense, Palaiseau, France) different target and catcher materials were investigated in terms of neutron yield. LULI consists of a Ti:Sapphire oscillator which is chirped[1] and amplified in a mixed Nd:silicate/phosphate-glass medium. With this method LULI is capable of delivering 30 J in 450 fs with a repetition rate of $8.\bar{3} \times 10^{-4}$ Hz, which corresponds to one shot in 20 min. The experiments were carried out

[1] Chirped pulse amplification (CPA) is a technique for amplifying an ultrashort laser pulse up to PW-intensities. The laser pulse is stretched temporally and spectrally with optical gratings to decrease the energy density prior to a second amplification. Afterwards the broad laser pulse is compressed again. Without *chirping* the pulse, the amplifier medium would be destroyed by non-linear optical effects.

at an energy of 20 J per pulse to preserve the compressor gratings, thus achieving a peak intensity of 7×10^{19} W/cm^2. As the target material CD^2 (thickness 5 mm), aluminum (30 μm), hydrated and deuterated titanium (50 μm) and gold (20 μm) were used. If present, CD^2, Ti:D, aerogel, or beryllium were used as catcher materials. Additionally, methods were applied to free the back of deuterated targets from protons in order to improve the acceleration effect for the deuterons. The best results were achieved with a gold target and a beryllium catcher. A neutron yield of 8×10^7 n/shot was achieved by the ^{9}Be(p,n)^9 reaction. This corresponds to 5×10^7 n/J, which is two orders of magnitude above the specific yield at ATLAS-10.

It is to be expected that the neutron yield with BOA-based ion acceleration can be significantly increased further. Experiments were carried out by [15, 17, 18] at the Trident laser facility of the Los Alamos National Laboratory. Since a BOA-based ion source was conceived for the compact HBS project, these experiments are explained in detail in Sect. 3.1.

2.1.2 Conventional Particle Accelerators

There are many different types of conventional accelerators which are designed for different applications according to their operating parameters (for example, particle type, energy, current, and duty cycle). Linear accelerators are of particular interest, which are described in more detail hereafter.

Ion Source The ion source is one of the most important components of an accelerator, since it generates the initial beam of the machine. In many types of accelerators, the ion source thus determines the pulse structure and the maximum particle current.

In principle ion sources consist of two main components: the plasma generator and the extraction system. The plasma can be generated in different ways, e.g., electrical discharges, heating, lasers, or by beams of other particles. In each and every case the plasma must be stable for long enough to extract a beam. The extraction system must be able to produce a beam of the correct shape and divergence to the next phase of the accelerator by extracting the correct ions and suppressing any unwanted ions, electrons, or neutral particles. The most important designs of sources for positive ions are electron bombardment sources, (Duo-) Plasmatrons, microwave ion sources, and laser ion sources [19, 20].

RFQ The high-frequency quadrupole resonator, usually referred to as an RFQ (radio frequency quadrupole) accelerator [21], is suitable for particle velocities of up to ~12% of the speed of light which corresponds to an energy of 7 MeV for protons. The four electrodes of the electrical quadrupole are arranged inside a cavity resonantor symmetrically around the particle beam. They are shaped such that the distance from the beam axis varies in the form of waves. As a result, the electrical alternating field of the stationary wave receives a longitudinal component, which alternately points into the beam direction and towards it. A continuous particle beam is thereby separated into particle packets during acceleration. The transverse components of the field acts focussing on the beam. RFQ resonators are operated at frequencies up to 500 MHz [22, 23].

Cavities and Drift Tubes In order to accelerate ions to higher energies additional stages after the RFQ are needed. In most cases cavity resonators are suitable for this purpose. These are cylindrical-symmetrical components in which a stationary wave of the electrical field, usually with different modes, can form due to resonance. For this purpose, high-frequency power is fed to the cavities, which is usually produced by klystrons. In most cases the mode TM_{010} is preferred, where the accelerating electric field points along the cylinder axis and the magnetic field lines run concentrically around the cylinder axis [24].

By using superconducting cavities, the attenuation of the injected high-frequency power can be significantly reduced [25] so that larger acceleration voltages per unit distance up to 35 MV/m are possible [26]. This makes the machine considerably more compact, which can drastically reduce construction costs. However, it must be kept in mind that the high costs of the superconducting components can counteract these savings again.

Extending an RFQ accelerator with an additional stage of cavities and drift tubes is associated with a significant cost increase. However, increasing the energy within an existing accelerator stage (e.g., by adding more cavities) leads only to a moderate cost increase.

Pulsing Each alternating voltage accelerator can, in principle, only accelerate those particles which reach the acceleration path at a suitable phase position of the alternating voltage. Therefore, the particle beam is always divided into bunches, i.e., pulsed and not continuous. This micropulse is, however, usually not referred to as pulsing in practice. As long as the high-frequency source operates continuously and an individual particle bunch is accelerated with each individual wave, continuous wave (cw) mode or continuous wave operation is used. Pulsed or pulsed beams are only referred to when the high-frequency voltage is not constantly present, but is switched on and off repeatedly, so that macropulses are generated [24].

Focusing In many linear accelerator structures, the particle beam must be focused along its path by special measures. For this purpose, quadrupole magnets and sextupole magnets are used, partly solenoids, through which the beam travels along its axis. The focusing elements are arranged alternately with the accelerating elements (RFQs or cavities).

The particles inevitably cause oscillations (so-called betatron vibrations) around their nominal path. The amplitude of these oscillations determines the diameter of the beam, thus the width of the magnetic pole shoes and thus the overall size and construction costs. The best focusing effect is achieved with alternating gradient focusing, also called strong focusing. The magnetic field of quadrupole magnets acts focusing on the beam of charged particles in the one transverse direction and defocussing in the other transverse direction. An arrangement in which multiple quadrupole magnets are rotated by 90° against each other, a strongly focusing net effect results. Illustratively, this corresponds to an arrangement of collecting and scattering lenses for light, which offers a focusing effect, too [24].

Further Accelerator Designs In addition to linear accelerators, other types of accelerators may also be suitable, for example cyclotrons and tandem accelerators.

A classical cyclotrons consists of a vacuum chamber in which two hollow electrodes with semicircular base are embedded. The vacuum chamber is located between the poles of a big electromagnet that creates a homogeneous and static magnetic field. The ion source is located in the center of the device and the ions are accelerated in the gap between the electrodes which are operated with alternating voltage. Due to the external magnetic field the particles are affected by Lorentz force and thus follow a spiral path whose radius increases with the particle energy. The classical cyclotron is only suitable for non-relativistic particles, since at relativistic velocities the particle's circulation period does not match the frequency of the accelerating voltage any more. By using an inhomogeneous magnetic field this mismatch can be balanced, so that final energies of 800 MeV can be achieved with protons. This inhomogeneous magnetic field acts defocusing on the particle bunches which is why additional focusing magnets have to be installed. These systems are called isochronous cyclotrons [24]. However, the maximum beam current is limited to a few hundred μA.

Tandem accelerators belong to the class of electrostatic accelerators and represent a further development of the Van-de-Graaff accelerators. The accelerating high voltage is generated by a Van-de-Graaff generator and can be as high as several MV, which is applied to the accelerator electrode, the terminal. The magnitude of the voltage is limited by the dielectric strength of the surrounding medium. For that reason the Van-de-Graaff generator and the ion beam tube are located inside a pressure vessel filled with an inert gas, e.g., nitrogen, carbon dioxide or sulfur hexafluoride SF_6. At Tandem accelerators the terminal is equipped with a stripping foil which removes the electrons from the previously negatively charged ions. Hence, the accelerating voltage can be applied again to the now positively charged ions. Final energies of up to 50 MeV are possible [27]. Here again, the maximum beam current is limited to a few hundred μA [28], so they are suitable for compact neutron sources such as the HBS prototype.

2.2 Ion-Target Interaction

In this section, the possible interaction mechanisms of the ions with the target nuclei are elucidated. In addition to the neutron-producing reactions relevant for CANS and the HBS in particular, these include competing processes, macroscopic effects that can affect the integrity of the target, and radiological aspects.

2.2.1 Neutron Producing Reactions

Musiol et al. [2], the reaction channels relevant for the generation of neutrons in CANS and laser-based neutron sources are explained in detail in this section. The neutron yield for different ion-target combinations can be estimated by considering the cross section $\sigma_{(\mathrm{a,n})}$ for the specific reaction channel $A(a,n)B$. For infinitesimal thin targets, consisting of a one-atomic layer, the neutron yield(rate) can be expressed as

$$Y = \dot{N}_{\mathrm{n}} = \sigma_{(\mathrm{a,n})}\, N_{\mathrm{S}}^{A}\, I_{\mathrm{a}} \quad , \tag{2.1}$$

where N_{S}^{A} describes the particle surface density of the target layer and I_{a} the incident particle current.

In order to maximize the neutron production rate $\dot{N}_{\mathrm{n}}$, thick targets are considered in which the incident particle beam is completely stopped. Therefore, it has to be taken into account that the reaction $A(a,n)B$ occurs in the whole energy range $[E_{\mathrm{a}}, 0]$ since the incident particles undergo energy loss due to ionization which is not involved in Eq. 2.1. For thick targets the yield is given as

$$Y = N_{\mathrm{V}}^{A}\, I_{\mathrm{a}} \int_{E_{\mathrm{a}}}^{0} \frac{\sigma_{(\mathrm{a,n})}(E)}{-\dfrac{\mathrm{d}E}{\mathrm{d}x}(E)}\, \mathrm{d}E \quad ,$$

whereby $-\frac{\mathrm{d}E}{\mathrm{d}x}$ denotes the specific stopping power due to ionization and N_{V}^{A} the particle density. If multiple reaction channels $i = 1, \ldots, n$ contribute to neutron production with ν_i neutrons each, e.g., $\nu_{(\mathrm{p,2n})} = 2$, the neutron yield reads

$$Y = N_{\mathrm{V}}^{A}\, I_{\mathrm{a}} \sum_{i=1}^{n} \nu_i \times \int_{E_{\mathrm{a}}}^{0} \frac{\sigma_{(\mathrm{a,nX})}^{i}(E)}{-\dfrac{\mathrm{d}E}{\mathrm{d}x}(E)}\, \mathrm{d}E \quad . \tag{2.2}$$

Since many reaction channels are opened at energies in the MeV range, in which partially charged reaction products are produced, which in turn can produce neutrons, the evaluation of the integral in Eq. 2.2 is not trivial. With the aid of Monte Carlo simulations, the yield can be precisely calculated using the databases for the possible reaction channels. In the following sections the neutron producing reaction channels for protons and deuterons and beryllium or lithium as target material as well as photonuclear neutron production in ^{238}U are discussed.

Compound Nuclear Reactions In compound nuclear reactions projectile and target nuclei combine to an intermediate nucleus that is in a highly excited state due to the gain in binding energy. After a typical lifetime between 10^{-19} s and 10^{-15} s the intermediate nucleus returns into the ground state by emission of secondary particles. The decay of the intermediate nucleus is completely independent of the nature of its formation. Compound nuclear reactions particularly occur at projectile energies which are clearly below the binding energy of a nucleon in the target nucleus [29, 30].

Reactions such as the induced nuclear fission, the evaporation phase of a spallation reaction and particle capture with gamma emission can also be understood as compound nuclear reactions. The deuterium-tritium fusion reaction used in neutron generators and fusion reactors also proceeds via a ^{5}He intermediate core state.

In most CANS beryllium or lithium are used as target materials which are usually impinged with protons or deuterons. Exemplarily, the cross sections for the most relevant reaction channels with protons impinging on beryllium and lithium and neutrons in the final state are shown in Fig. 2.3.

The reaction channels, shown as colored curves, differ in the final state, i.e., the type and number of secondary particles emitted during the reaction. Each reaction channel exhibits a specific threshold energy which essentially depends on the binding energy of the secondary particles in the nucleus. The more secondary particles appear in the final state, the higher is the threshold energy of the corresponding reaction channel. The black curve in the diagrams represents the weighted sum of the cross sections of all possible reaction channels with neutrons in the final state. This cross section is particularly interesting when the neutron yield is to be estimated. However, it must be kept in mind that competing processes lead to distortion. For example, the primary ions lose a significant amount of their energy due to ionization in the material before one of the neutron-producing reactions can take place. Furthermore, e.g., beryllium has a particularly high (n,2n) cross section so that the neutron yield is increased.

Direct Nuclear Reactions Direct nuclear reactions take place without the formation of a highly excited compound nucleus. Examples for direct nuclear reactions are inelastic scattering, in which kinetic energy of the projectiles is converted into excitation energy of the target nucleus, the first phase of a spallation reaction, where the target nucleus is shattered into several fragments, and knock-out reactions in which the projectile knocks out one or more nucleons from the target nucleus without formation of an intermediate nucleus. Charged particles can also interact with the Coulomb field of the target nucleus. The time scale of such reactions is significantly shorter than that of compound nuclear reactions and corresponds to the flight time of the the projectile along the nucleus' diameter. A typical value is 1×10^{-22} s for fast projectiles with an energy $\mathcal{O}(1\,\text{MeV})$. Secondary particles produced in direct nuclear reactions are usually strongly forward-directed due to the low energy and momentum transfer [2, 29, 30].

Considering deuterons impinging on beryllium nuclei several direct nuclear reactions are possible relevant to neutron production. In a Be(d,n) *stripping* reaction, the proton of the deuteron is transferred to the beryllium nucleus so that the remaining neutron can propagate almost unimpeded. In addition, it is possible that the deuteron decays into its constituents in the coulomb field of the beryllium nucleus, so that neutron and proton propagate separately from each other. This Be(d,pn) reaction is called *break-up*. Because of these additional reaction modes, the cross section of neutron production, cf. Fig. 2.4, is significantly larger for deuterons on beryllium than for protons. By comparing to the proton cross sections in Fig. 2.3 it is evident

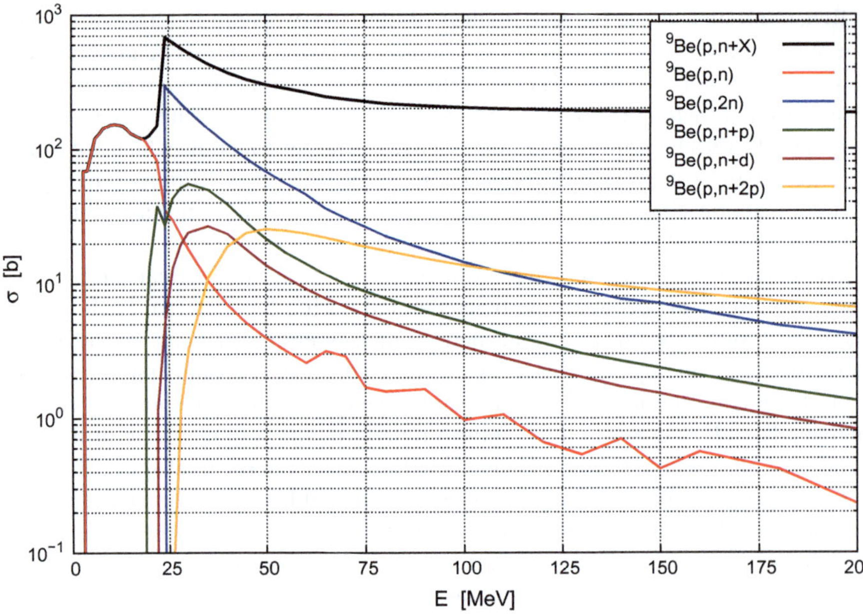

(a) Beryllium as target material.

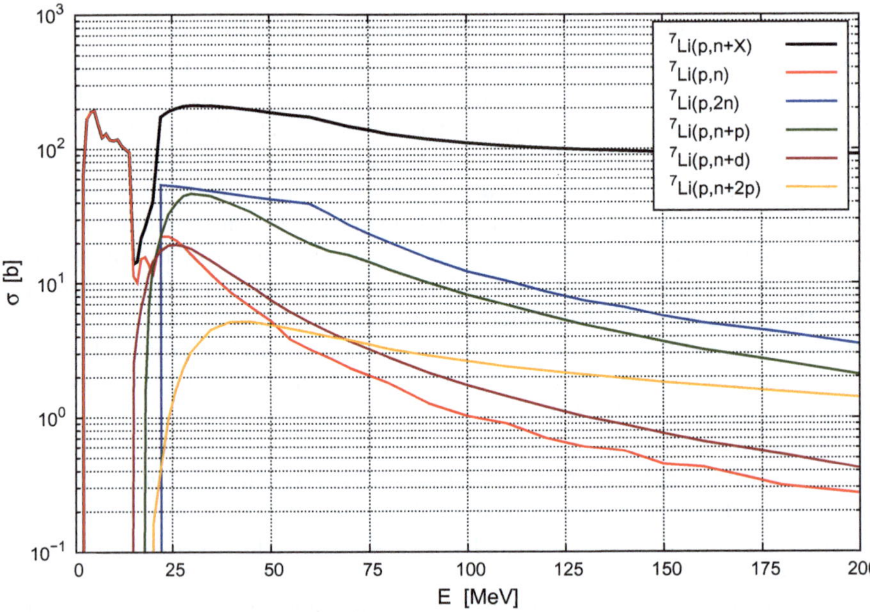

(b) Lithium as target material.

Fig. 2.3 Energy-dependent cross sections of neutron producing reactions with protons as projectiles. Data taken from the TENDL-2015 library [31], see Sect. 2.5.4

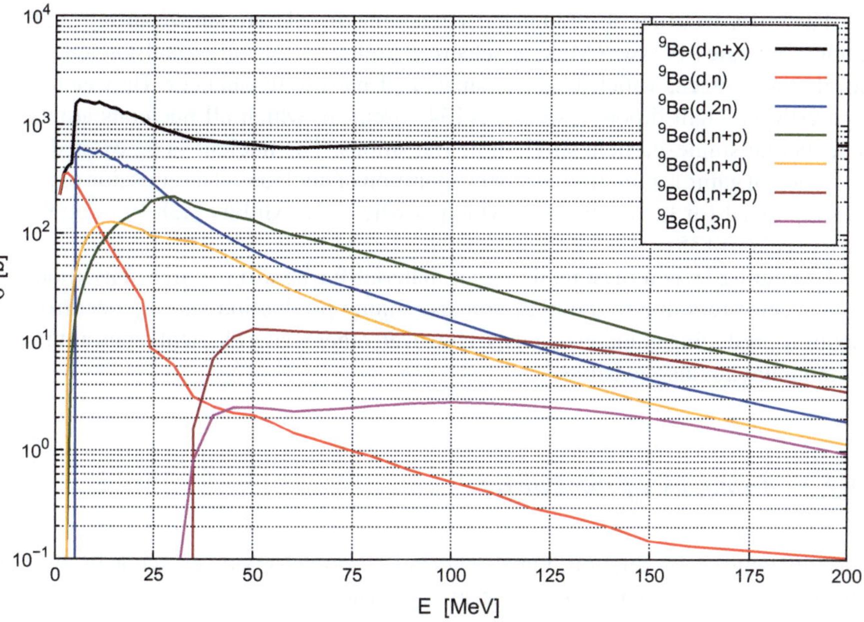

Fig. 2.4 Energy-dependent cross sections of neutron producing reactions for deuterons impinging on beryllium. Data taken from the TENDL-2015 library, see Sect. 2.5.4

that the stripping reaction results in an increased (d,n) and (d,2n) cross section and the breakup reaction leads to a markedly increased (d,n+p) cross section.

Photonuclear Reactions Gamma radiation of sufficiently high energy can interact with nuclei which are called photonuclear reactions. There are different reaction channels, including the emission of neutrons or fission. Such gamma radiation can be produced as bremsstrahlung by bombarding high-Z materials with high-energy electrons.

The corresponding cross section for ^{238}U is shown as a red curve in Fig. 2.5a. The emission of bremsstrahlung is the predominant effect for light, charged high-energy particles when interacting with matter. Accordingly, the underlying cross section is comparatively high. However, below an energy of approx. 10 MeV the main mechanism for energy loss is ionization [32]. The blue curve in Fig. 2.5a shows the spectrum of the bremsstrahlung photons generated by 50 MeV electrons in ^{238}U. The spectrum exhibits a continuous component which behaves like $E - E^2$ according to Kramer's law [33]. In the X-ray region ($E_\gamma \lesssim 500\,\text{keV}$), the X-ray continuum is proportional to E and superimposed by various effects. These include, in particular, the discrete characteristic radiation which arises due to transitions between the energy levels of the inner electron shells, as well as its absorption bands. In the high-energy range, the spectrum follows a continuous $-E^2$ curve. The maximum energy of the bremsstrahlung photons corresponds to the kinetic energy of the incident electrons, in this example 50 MeV.

The cross sections of photonuclear reactions in ^{23}U are shown in Fig. 2.5b. For photon energies below 8 MeV, a selective excitation of individual nuclear energy levels takes place, which leads to small and narrow resonances. Between 8 MeV and 20 MeV giant dipole resonances [34] occur, in which all nucleons are excited to coherent single-particle single-hole transitions. The nucleus is de-excited by the emission of gamma radiation and/or evaporation of nucleons. The emission of protons is suppressed for heavy nuclides which is why ^{238}U exhibits high (γ,n) and (γ,2n) cross sections [35].

Uranium-238 is a reasonable target material for neutron sources based on electron accelerators due to the high bremsstrahlung production and the occurrence of neutron-producing giant resonances. However, other heavy elements such as tungsten or tantalum are also feasible.

2.2.2 Integrity of the Target

Despite neutron production, many other microscopic processes occur in the target due to the bombardment of high energy ions. These microscopic processes can have a significant effect on the macroscopic characteristics of the materials used and therefore must not be neglected in the design of the target. In some configurations, the target is also intended to serve as the end window of the accelerator vacuum, as a result of which the requirements for the target material once again increase considerably. In this section solid targets are considered. But for liquid targets, the same questions arise for the wall material and additional issues concerning the fluid dynamics of the liquid target jet.

Hydrogen Embrittlement Apart from the pure thermal load of the target, an additional problem arises from the bombardment with protons or deuterons. Because the ions in the material can be stopped, they can recombine with electrons, so that hydrogen is generated. If the hydrogen can not diffuse fast enough to the surface, bubbles can form which make the target increasingly brittle until material failure occurs. This phenomenon is referred to as hydrogen embrittlement or blistering.

Beryllium as a common target material is strongly affected by this phenomenon. A common practice against this problem is the use of a backing plate of a material which has favorable properties with respect to the deposition and diffusion of hydrogen. In this approach the fact is utilized that the stopping power for protons and heavier ions leads to the characteristic Bragg curve [32], which indicates that most of the incident ions are deposited at the maximum penetration depth (i.e., the *Bragg peak*) in the material. For the *Riken Accelerator-Driven Neutron Source* (RANS) at Riken University, Japan [36], a PE-moderated neutron source driven by a 7 MeV proton linac with a maximum current of 100 μA, a target was developed and patented [37, 38] at which the irradiated beryllium layer is bonded on a backing plate of vanadium which offers a reasonable heat conductivity and resistance to hydrogen embrittlement. The beryllium layer thickness is chosen so that the proton's Bragg peak is located in

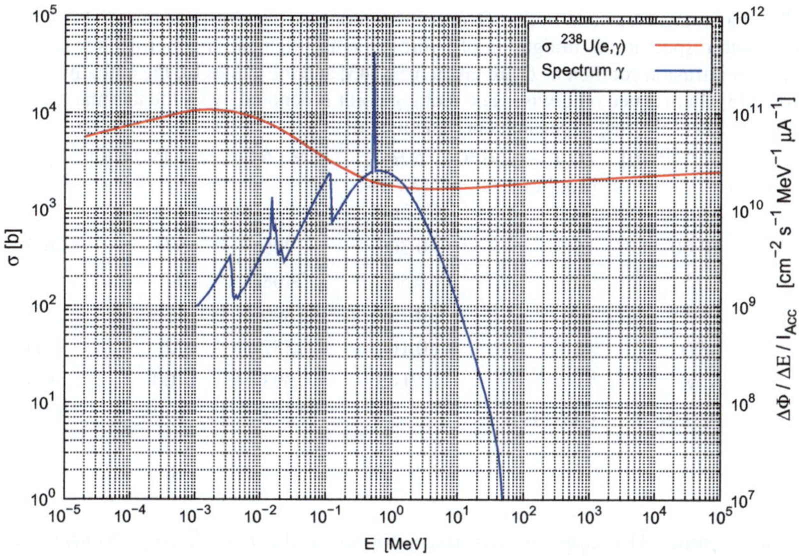

(a) Cross section for bremsstrahlung emission of electrons in ^{238}U taken from the ENDF/B-VII.1 data library [45] (red curve and left scale) and spectrum of bremsstrahlung photons produced by 50 MeV electrons calculated with MCNP6.1.1b (blue curve and right scale).

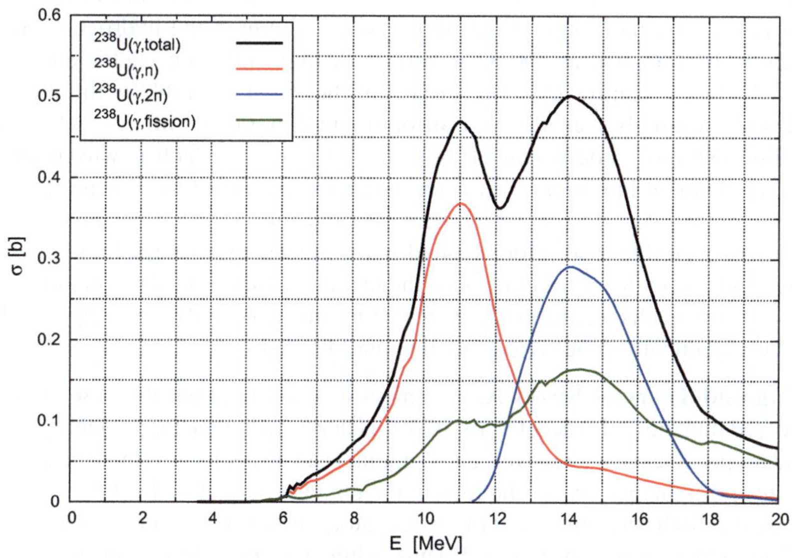

(b) Photonuclear cross sections for ^{238}U taken from the ENDF/B-VII.1 data library.

Fig. 2.5 Cross sections and spectra relevant to photonuclear neutron production with electrons impinging on ^{238}U

Table 2.1 Parameters relevant for the selection of appropriate backing plate materials for the target assembly with respect to the mitigation of hydrogen embrittlement and thermal stress. Values of the hydrogen diffusion coefficient D are taken from [43] and the values for the enthalpy of solution ΔH from [43] if not stated otherwise. The melting point is denoted by Θ, the thermal conductivity by κ, and the linear thermal expansion coefficient by α_l

	$\sigma_{(n,\gamma)}^{\text{thermal}}$ [b]	D ($T = 200\,°C$) [$10^{-6}\,cm^2/s$]	ΔH [eV/H-atom]	Θ [°C]	κ [W/m K]	α_l [10^{-6}/K]
Ta	20.00	15	0.30	2996	57	6.5
Nb	2.15	30	0.35	2468	54	7.1
V	5.08	100	0.25	1890	31	8.3
Pd	6.90	10	0.10	1552	72	11.0
Cu	3.78	1	–0.45	1093	400	16.8
Be	< 0.01	0.05 [44]		1278	190	12.3

the backing plate. This approach is also adopted at the *Low Energy Neutron Source* (LENS) at the Indiana University, USA [39], the iBNCT project at the University of Tsukuba, Japan [40], and will also be used for the design of the HBS target.

Table 2.1 lists the most important parameters for tantalum, niobium, vanadium, palladium, copper, and beryllium [41]. Beryllium is also considered to show the differences to a bulk target without backing plate. In order to minimize the impact on the neutron yield, the material should have a small absorption cross section, which is listed in the first column. In the second column the hydrogen diffusion coefficient D evaluated at 200 °C is shown. A high diffusion coefficient ensures that the hydrogen can travel to the surface of the target assembly without forming bubbles. Accordingly, vanadium is an optimum candidate with $D = 10^{-4}\,cm^2/s$. Beryllium exhibits a coefficient of $D = 5 \times 10^{-8}\,cm^2/s$, which is why it tends to blistering. Another important parameter in the assessment of hydrogen diffusion is the enthalpy of solution ΔH, which describes the energy release when a hydrogen atom is absorbed (the absorption of a hydrogen molecule is twice as high). Since the energy stems from the crystal lattice of the metal, endothermic absorption reactions ($\Delta H < 0$) are much more probable than exothermic ($\Delta H > 0$). In this light niobium is the best candidate while copper is ruled out [42, 43].

Heating and Thermal Stress As a conservative assumption, the case should be considered that the entire accelerator power is deposited in the target. Dissipation by secondary particles (e.g., neutrons and gammas) can be neglected since the dominant mechanism for energy loss of the ions in the target is ionization [32].

Due to the high performance of the accelerator, surface power densities in the range 1 kW/cm^2 and above can be reached. Thus, while designing the cooling system, care must be taken not to exceed the melting point of the target material, see Table 2.1, but also to keep the thermal stresses within the tolerable range over the full depth of the active target layer up to the cooling channels.

Ansys simulations conducted by [45] for beryllium target assemblies with a vanadium backing plate show that for accelerator powers of 5, 10, 20, 30, and 50 kW the target temperatures are far below the melting point. In Fig. 2.6b the simulated one-dimensional temperature distribution in a generic target assembly is shown for the aforementioned accelerator powers and an energy of 50 MeV. According to the geometry depicted in Fig. 2.6a the vanadium backing extends from –10 to 0 mm with a cooling channel diameter between –6 mm and –1 mm. The beryllium layer begins at 0 mm. Due to the lower penetration depth of heavy ions, the thickness of the beryllium layer in the case of deuterons amounts to 7 mm and in the case of protons 14 mm. The temperatures at the bombarded surface for 5 kW protons are 103 °C and for deuterons 98 °C. As the accelerator performance increases, the difference between the maximum temperatures of protons and deuterons significantly increases due to the spatial distance of the bombarded surface from the cooling channel. Thus, at 50 kW, the maximum temperature for protons is 985 °C and for deuterons 848 °C. The temperature slowly decreases up to the interface, then decreases rapidly in the vanadium to a minimum of 22 °C in the coolant. This steep temperature gradient can cause high stresses, which can even lead to material failure.

In the case of targets with backing plate, the different coefficients of thermal expansion of the materials must also be taken into account, see Table 2.1. For strongly different coefficients of expansion an insulating gap forms, which prevents heat dissipation from beryllium, if the materials are not bonded. For bonded target layers different expansion coefficients will lead to high stresses at the interface. For a pulsed operation with low duty cycles, it must also be taken into account that significant periodic temperature fluctuations occur, which impair the material integrity in the long term.

DPA Production An important parameter for assessing the integrity of irradiated crystalline and mineral components, especially metals, is the number of *displacements per atom* (DPA). This quantity describes how many atoms have been displaced from their initial position in the crystal lattice due to interactions with high energy particles, e.g., neutrons, ions, electrons and photons, normalized to the total number of atoms. Due to the thermal motion of the atoms, it is possible that lattice defects can disappear by recombination processes which is called *annealing*. The annealing rate increases with the temperature of the material. Typical maximum values, up to which components can still be regarded intact, are in the single- or double-digit DPA range, depending on the material and its intended purpose [46, 47].

The number of displacements can be derived from the NRT model [48] which provides an empirical correlation between the total number of displacements N_{NRT} and the total energy E_a of the *Primary Knock-on Atom* (PKA)[2]:

$$N_{\mathrm{NRT}} = 0.8 \times \frac{E_a\,[\mathrm{MeV}]}{2\,E_{\mathrm{d}}\,[\mathrm{eV}]} \quad , \tag{2.3}$$

[2] A Primary Knock-on Atom is an atom that is the first atom that is displaced from its lattice position by irradiation. With sufficient energy, the PKA can induce subsequent displacements of other atoms resulting in a *damage cascade*.

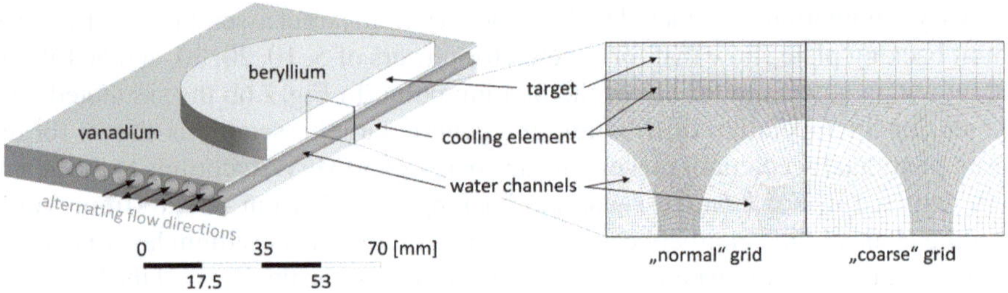

(a) Left side: geometry of the target assembly, right: selection of a cut through the grids used for the finite-element-method. [58]

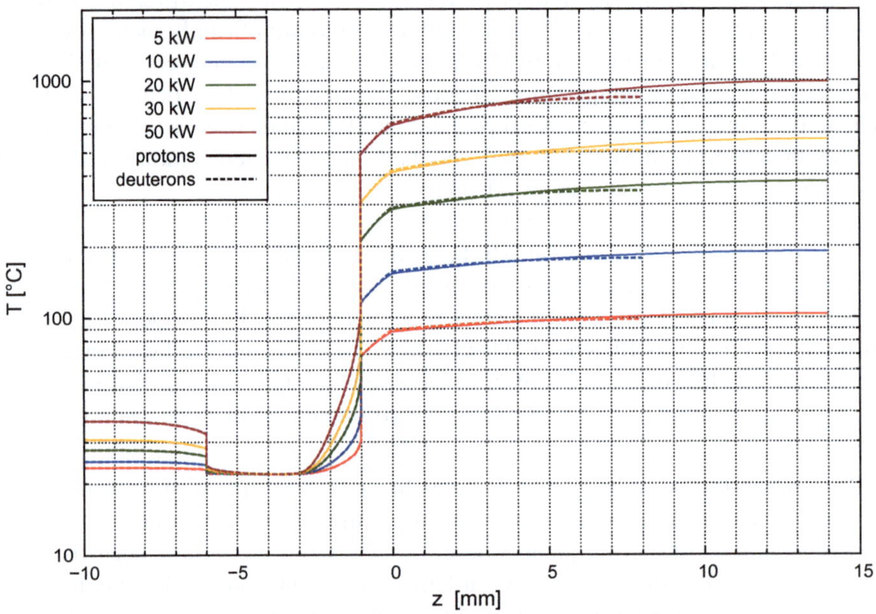

(b) Calculation results for protons and deuterons at different accelerator powers. Data is taken from [58].

Fig. 2.6 Ansys geometry and calculation results of the one-dimensional temperature distribution in a generic beryllium target assembly with a vanadium backing plate and a cooling channel bombarded with protons or deuterons at different accelerator powers. The vanadium backing extends from −10 to 0 mm with a cooling channel diameter between −6 mm and −1 mm. The beryllium layer begins at 0 mm

where E_{d} denotes the average threshold displacement energy for the PKA in the material which amounts to 31 eV for beryllium [49]. The value of E_a is calculated by MCNP for each history.

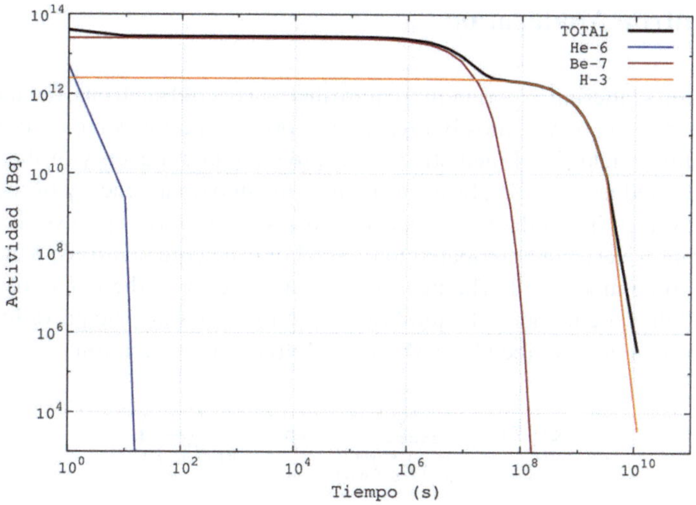

Fig. 2.7 Total and specific activity of ^{6}He, ^{7}Be, and ^{3}H in the beryllium target after 2000 h of irradiation with a 50 MeV, 2.25 mA proton beam [51]

2.2.3 *Activation of the Target*

Stable nuclides can be transmuted into radioactive nuclides by bombardment with, e.g., neutrons or ions. Depending on the type of the incident particles and the irradiated nuclides, different reaction channels exist which lead to different products. In nuclear facilities activation is basically a result of the intense neutron fluxes. For the target of an accelerator-driven neutron source, however, the main source for activation is the irradiation with the high-energy primary ions, whereby the most important reaction channel for beryllium-targets is ^{9}Be(p,^{3}H)^{7}Be, in which tritium and ^{7}Be are produced. This reaction has a threshold energy of 13.4 MeV [50]. Another reaction channel is ^{9}Be(n,3n)^{7}Be with a rather high threshold energy of 25 MeV [31] and is established by the high-energy neutrons produced in the target.

In [51], the production of tritium and ^{7}Be in a beryllium target was calculated for a 50 MeV proton beam with a mean current of 2.25 mA. The time dependence of the activity after 2000 h of irradiation can be seen in Fig. 2.7. It is evident that the short term activity is governed by ^{7}Be, whereas tritium dominates after approximately 8 months (2×10^{7} s). Contributions from activated impurities in the beryllium target can be neglected. Since no information on total mass of the irradiated material is available, a comparison with the clearance levels of the German Radiation Protection Ordinance (StrlSchV) is not possible. It can, however, be noted that the exemption values for tritium (10^{9}Bq) and ^{7}Be (10^{7}Bq) are clearly exceeded, so that the irradiated target is subject to surveillance according to the German Atomic Energy Act.

2.3 Neutron Moderation

Most neutron scattering experiments require thermal or cold neutrons. For the production of neutrons, however, mostly high-energy primary particles are used to achieve a high neutron yield. For kinematic reasons, the maximum energy of the spectrum of the generated neutrons is always as high as the maximum energy of the primary particles, even if the production of such neutrons is strongly suppressed. However, the spectrum is very hard compared to thermal or even cold neutrons, which is why they have to be moderated. The neutrons transfer energy to the nuclei of the moderator until they are in thermal equilibrium with the material. The prevailing energy distribution can be described by a Maxwell-Boltzmann distribution

$$\Phi(E) = \frac{2\sqrt{E}}{\sqrt{\pi\, k_{\mathrm{B}}^3\, T_{\mathrm{mod}}^3}} \exp\left(-\frac{E}{k_{\mathrm{B}}\, T_{\mathrm{mod}}}\right) \tag{2.4}$$

under negligence of energy-dependent absorption processes. The most probable energy of the neutrons corresponds to the ambient temperature T_{mod}. Since there is always a weak absorption, which, according to the $1/v$-law preferably concerns neutrons of the lowest energy, the effective temperature of the so-called *neutron gas* is slightly higher than the temperature of the moderator. In the following sections the basic principles of moderation and relevant parameters for typical thermal and cold moderators are examined.

2.3.1 *Thermalization*

The neutrons are decelerated mainly by elastic scattering with the moderator nuclei. The energy transfer ΔE of a neutron to a moderator nucleus can be calculated using the classical collision kinematics and yields

$$\Delta E = E - E' = E\,\frac{2\,A}{(A+1)^2}\,(1 - \cos\theta) \quad , \tag{2.5}$$

where E and E' denote the initial and final neutron energy, respectively, A the mass number of the target nucleus and θ the scattering angle in the center-of-momentum frame. For the central impact ($\cos\theta = 1$) the energy transfer is therefore at maximum. However, a large number of collisions have to be averaged so that the average cosine of the scattering angle must be considered, which can be expressed by transformation into the laboratory system $\hat{\theta}$ as

$$\overline{\cos\hat{\theta}} = \frac{1}{2}\int_{-1}^{+1} \frac{A\,\cos\theta + 1}{\sqrt{A^2 + 2\,A\,\cos\theta + 1}}\,\mathrm{d}\cos\theta = \frac{2}{3\,A} \quad . \tag{2.6}$$

Table 2.2 Relevant parameters for estimating the performance of typical moderator materials. The slowing-down power $\xi\ \Sigma_{el}$ and the moderating ratio $\xi\ \frac{\Sigma_{el}}{\Sigma_{abs}}$ are evaluated for thermal neutrons ($E = 25.3$ meV). The moderating time t_{mod} and the mean number $\bar{n}_{mod}$ of collisions needed for thermalization are related to an initial energy of $E = 2$ MeV. The diffusion time t_{diff} and the mean number $\bar{n}_{diff}$ of collisions during diffusion are given for thermal neutrons [52]

	$\xi\ \Sigma_{el}$ [cm^{-1}]	$\xi\ \frac{\Sigma_{el}}{\Sigma_{abs}}$	t_{mod} [μs]	$\bar{n}_{mod}$	L [cm]	t_{diff} [s]	$\bar{n}_{diff}$
H_2O	1.36	62	10	20	2.7	2.1×10^{-4}	117
D_2O	0.18	5000	46	27	170.0	1.5×10^{-1}	16042
Be	0.16	145	67	86	21.0	4.3×10^{-3}	806
C	0.06	165	150	114	54.0	1.2×10^{-2}	1056

When substituting $\cos\theta \rightarrow \overline{\cos\hat{\theta}}$ it is evident from Eqs. 2.5 and 2.6 that the energy transfer to light nuclides is maximal [52].

In order to estimate the quality of thermal and cold moderators, not only the magnitude of the energy loss per collision but also the collision probability must be considered. Hence, the product of the mean logarithmic energy decrement, which is a measure of the mean energy loss per collision but at the same time is independent of the incident neutron energy and therefore a material constant, and the macroscopic elastic scattering cross section are considered as the slowing-down power $\xi\ \Sigma_{el}$. Since a high absorption cross section Σ_{abs} has a negative impact on the performance of a moderator, the moderation ratio $\xi\ \frac{\Sigma_{el}}{\Sigma_{abs}}$ should also be considered. The cross sections for elastic scattering and radiative capture for light and heavy water, beryllium, and natural carbon are depicted in Fig. A.1 in the appendix.

For these thermal moderators, the values of the moderator performance parameters are listed in Table 2.2. Since the cross sections are energy-dependent, the slowing-down power and the moderating ratio are given for thermal neutron energies. Additionally, the time t_{mod} needed for a 2 MeV[3] neutron to be thermalized and the mean number of collisions during that time is indicated. Regarding parasitic reactions, basically absorption, the diffusion length L of thermalized neutrons, the associated diffusion time t_{diff} to thermalized neutrons, and the number $\bar{n}_{diff}$ of collisions during this time is also listed.

Some qualitative statements about the performance of the moderators can already be made by means of these values. As can be seen from Eq. 2.5, hydrogen-based moderators exhibit an excellent slowing-down power. Therefore, only few collisions are necessary for the moderation of a 2 MeV neutron. Due to the relatively large absorption cross section, however, the moderating ratio is small, which is why the diffusion time of thermal neutrons is very short. Heavy water offers a significantly lower slowing-down power but a very high moderating ratio due to the very low absorption cross section, cf. Fig. A.1 in the appendix. Beryllium has a similar slowing-down power as heavy water but also a higher absorption cross section which is why

[3]This is the mean energy of neutrons produced in nuclear fission of ^{235}U.

the moderating ratio is only slightly better than for water. Graphite has a similar moderating ratio as beryllium, since the slowing-down power and the absorption cross sections are significantly smaller.

On the basis of these statements it can be assumed that heavy water, beryllium and graphite are more suitable for large volume moderators, such as in nuclear reactors, or as reflectors. Compact moderators, on the other hand, are based on hydrogen.

2.3.2 *Production of Cold Neutrons*

Physical Principles As indicated by Eq. 2.4, cryogenic moderators are necessary to produce cold neutrons. Since cryogenic moderators usually occur as liquids or solids, further effects must be taken into account in addition to the elastic scattering of individual particles. These effects occur due to the de Broglie wavelength of low-energy neutrons being in the same range as the atomic bonds in the molecules or crystal lattices. In order to physically describe these phenomenons the scattering cross section is decomposed into a coherent part and an incoherent part as $\sigma = \sigma_{\mathrm{coh}} + \sigma_{\mathrm{inc}}$. *Coherent scattering* takes into account interference effects of the scattered neutrons and is therefore of particular importance for crystalline solids like graphite or beryllium, whereas *incoherent scattering* describes non-interfering final states, being of importance for hydrogenous solids like polyethylene or solid methane. The coherent and incoherent cross section are further decomposed into elastic and inelastic parts. The elastic scattering takes place with the entire crystal lattice. Due to the very high effective mass of the lattice compared to the neutron the energy transfer is negligible. Inelastic scattering results in an energy loss for the neutron with an excitation of the target which may correspond to the production of phonons in crystalline materials, the excitation of rotational or vibrational degrees of freedom in molecules, or molecular recoil motions in liquids.

The cross section for low-energy neutrons in gases, liquids and solids can be written as [53]

$$\begin{aligned} \sigma &= \frac{\sigma_{\mathrm{free}}}{2\,k_{\mathrm{B}}\,T}\sqrt{\frac{E'}{E}}\,\mathrm{e}^{-\beta/2}\,S(\alpha,\beta) \quad \text{, with} \\ \alpha &= \frac{E' + E - 2\,\mu\,\sqrt{E\,E'}}{A\,k_{\mathrm{B}}\,T} \quad \text{and} \\ \beta &= \frac{E' - E}{k_{\mathrm{B}}\,T} \quad , \end{aligned} \tag{2.7}$$

which encompasses a scattering function $S(\alpha, \beta)$, where α and β describe the momentum and energy transfer, respectively, σ_{free} the free scattering cross section of the material composition neglecting lattice and molecular effects, μ the cosine of the scattering angle, and A the ratio of the scatterer mass to the neutron mass.

The scattering function has a specific shape depending on the material and the reaction. For inelastic processes, for example, a series expansion according to phonon modes is performed as given in Eq. A.3 in the appendix. Rotational and vibrational modes, on the other hand, are treated as discrete oscillators, with a diffusion term additionally being introduced for liquids, see Eq. A.2. Incoherent elastic scattering, as given by Eq. A.4, is important for hydrogenous solids and results essentially from the zero-th phonon-mode of the phonon expansion for inelastic processes and thus has no energy transfer but a finite interaction probability. In the case of coherent elastic scattering in crystalline solids, given by Eq. A.5, an expansion according to Bragg edges and associated crystallographic structural factors is carried out. The series expansions of $S(\alpha, \beta)$ is an approximation which assumes that the spins of the target material particles are distributed randomly. As this is not true for materials containing molecules of hydrogen or deuterium, an explicit quantum-mechanical formula is required [54, 55].

The scattering cross sections containing $S(\alpha, \beta)$ corrections for common cryogenic moderator substances are depicted in Fig. A.2 in the appendix.

Cold Moderator Properties Ideally cold moderators are fed with thermal neutrons to increase the effectiveness due to a compact design. The conclusions made in Sect. A.1 can be transferred to cryogenic moderators, which is evident when considering the cross sections (A.1) in the thermal and cold energy range. However, at temperatures around 20 K pure hydrogen or deuterium, organic compounds such as methane or aromatic compounds such as mesitylene or xylene are used.

According to the experiences made by, e.g., [56–58], solid methane offers an excellent yield of cold neutrons. However, it has a poor radiation resistance due to its crystalline structure [59, 60]. As a result of the irradiation, lattice defects are induced, which represent an energetically higher state. These lattice defects can recombine exothermically initiated by thermal energy input, which is called *annealing*. If the stored lattice energy is sufficiently high, even a small heat input due to irradiation with neutrons can trigger an avalanche-like annealing, so that abruptly a large amount of heat is produced that can destroy the moderator and other components. This phenomenon is well known from nuclear graphite in reactor technology under the term *Wigner energy*. In order to prevent spontaneous and sudden annealing, methane moderators must be heated and re-frozen at regular intervals.

Studies on aromatic compounds have shown that similar cold neutron yields can be achieved as with methane, while mesitylene, xylene or toluene are much more resistant to radiation than methane. With those compounds, the advantages of solid cryogenic moderators can be exploited without having to frequently carry out manual annealing [61, 62].

However, pure hydrogen moderators are still of great relevance since in recent years interesting properties have been discovered which can be exploited in order to significantly increase the brilliance of cold neutrons. The H_2 molecule is subdivided into ortho- and para-states according the two possible triplet and singlet nuclear spin states. In Fig. 2.8a the total scattering cross sections of both hydrogen spin configurations are depicted. The drop of the parahydrogen's scattering cross section below

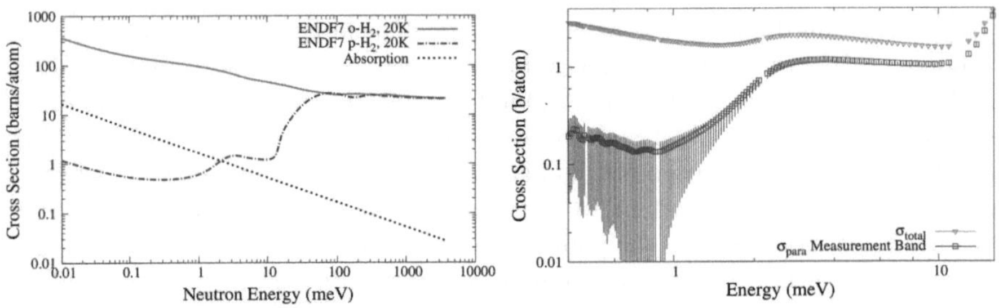

(a) Para- and orthohydrogen scattering cross sections at 20 K and the absorption cross section from ENDF/B-VII.

(b) Total cross section (triangles) of hydrogen and parahydrogen scattering cross section (squares) measured by [77] at 15.6 K. The upper error bar on the parahydrogen cross section is an estimated systematic error and the lower error bar is given by the upper limit on the orthohydrogen contamination.

Fig. 2.8 Ortho- and parahydrogen cross sections. Recent measurements indicate a inconsistency in the ENDF/B-VII.1 data for parahydrogen. It appears that the parahydrogen represented by this dataset is contaminated with a small amount of orthohydrogen [64]

0.05 eV is due to intermolecular interference whereas the drop at 0.003 eV results from spin coherence scattering. These effects are frozen below the corresponding energies in contrast to orthohydrogen. Thus, the shape of the parahydrogen scattering cross section implies that the mean free path $L_t = 1$ cm of thermal neutrons is much shorter than the mean free path $L_c = 11$ cm of cold neutrons [63].

The thermal neutrons entering the cold moderator from the thermal moderator are scattered down into the cold energy region after few collisions. Due to the larger mean free path, the cold neutrons can leave the moderator without further collisions. The density of the cold neutrons is therefore higher within a layer with thickness about L_t near the moderator wall. Consequently, the optimum shape is close to a tube of length L_c and radius L_t inserted into the thermal moderator. Thus, the cold moderator is fed from all sides and cold neutrons which do not leave the parahydrogen in forward direction can be scattered back by the surrounding thermal moderator. A large increase in cold neutron brilliance is reported by [65, 66] by exploiting the properties of low-dimensional parahydrogen moderators.

However, recent measurements on the scattering cross section of parahydrogen conducted by [64] indicate that the cross section data from ENDF is inconsistent in a sense that the underlying parahydrogen was contaminated with a small amount of orthohydrogen shifting the scattering cross section to higher values significantly. This becomes evident by comparing the lower curves of the Fig. 2.8a and b. The large lower error bars of the remeasured scattering cross section, which stem from the upper limit on the uncertainty of a possible orthohydrogen contamination, indicate that L_c could be even longer for 100% pure parahydrogen.

2.4 Neutron Instrumentation

Components of the instrumentation, the experimental setup behind the moderator, which are relevant for designing and optimizing (for example with respect to pulse structure and brilliance) the moderator system of the HBS, are described in this section.

Neutron Guides [67, 68] After moderation the angular distribution of the neutrons is almost isotropic. Hence, the flux at the experiment behaves like r^{-2} with the distance of the experiment to the moderator. Neutron guides can be used to increase the flux at the experiment by transporting the neutrons from the moderator with a larger divergence and focus them on the sample. This can be understood by considering the wave-particle duality which implies that neutrons can be described as particle waves and thus follow the same laws for refraction and reflection as electromagnetic waves. A neutron guide consists of four neutron mirrors forming a tube along the flight path of the neutrons. The lower index of refraction, $n_i < n_0$, of the neutron mirror material compared to vacuum gives rise to a critical angle of total reflection θ_{crit} which can be derived by assuming an effective interaction potential referred to as the Fermi Pseudopotential [69]

$$V(\vec{r}) = \sum_i \frac{2\,\pi\,\hbar^2}{m}\, b_{\text{c}}\, \delta^3(\vec{r} - \vec{r}_i) \approx \frac{2\,\pi\,\hbar^2}{m}\, b_{\text{c}}\, \rho_N \tag{2.8}$$

which results from treating the incident neutrons as particle waves and applying a partial wave decomposition for the time-dependent Schrödinger equation. Hereby ρ_N describes the particle density and b_{c} the coherent scattering length. The critical angle of refraction is derived by demanding the neutrons kinetic energy $E_\perp$ of the perpendicular component of the wave number $k_\perp$ to be lower as the interaction potential from Eq. 2.8:

$$\begin{aligned} E_\perp &= \frac{\hbar^2\, k_\perp^2}{2\,m} = \frac{2\,\pi^2\,\hbar^2\,\sin^2\theta}{m\,\lambda^2} \overset{!}{<} V \\ \Rightarrow \theta_{\text{crit}} &\approx \sin\theta_{\text{crit}} = \lambda\sqrt{\frac{\rho_N\, b_{\text{c}}}{\pi}}\,, \end{aligned} \tag{2.9}$$

whereby the critical angle is measured tangentially to the surface and implies that all neutrons hitting the mirror at an angle $\leq \theta_{\text{crit}}$ are totally reflected and transported by the neutron guide. The largest critical angle is achieved with nickel and can be approximated by $\theta_{\text{crit}}^{\text{Ni}}/\lambda = 0.1°/\mathring{A}$.

By using alternating layers of materials with different indices of refraction like Ni and Ti or pure isotopes of ^{58}Ni and ^{62}Ni and decreasing thickness, critical angles of a few degree for cold and $\leq 1°$ even for thermal neutrons are possible. Such systems are called supermirrors and typically consist of several hundred layers. The critical angle for a supermirror is given as $\theta_{\text{crit}}^{\text{mirror}} = m \cdot \theta_{\text{crit}}^{\text{Ni}}$ according to Eq. 2.9.

On the basis of these considerations, it is clear that during the development of the moderator system particular attention should be payed to maximize the brilliance at the moderator surface or the neutron entry point in the instrumentation, respectively.

Choppers and Energy Selectors [68] For time-of-flight measurements a determination of the energy is only possible if the source of the neutrons is pulsed. In the case of continuous sources or pulsed sources whose duty cycle and pulse lengths are not suitable for the experiment, choppers are used which divide the flux of neutrons into the correct pulse structure.

This is a rotating disk made of absorbing material which is equipped with transparent windows or openings. By using two choppers arranged one behind the other, an energy range of the neutrons can be selected by adjusting the rate of rotation and the distance. The same effect can also be obtained when the chopper is designed as a thick, rotating wheel, in whose lateral surface helical grooves are implemented. Only the neutrons with a speed adapted to the speed of the chopper can pass through the groove without being absorbed by the chopper material. These choppers are called *velocity selectors* and provide a relatively narrow band of neutron energies.

Since the use of choppers for the correction of the pulse structure always reduces the neutron flux at the sample, during the design of the neutron source it should be taken into account which pulse structures and repetition rates are useful for the experiments. For increasing pulse length the number of neutrons lost due to pulse shaping is also increasing. For high repetition rates the faster neutrons of the selected energy range can overlap at the detector with the slower neutrons of a preceding pulse. This effect is called *frame overlap* and requires further pulse shaping reducing the amount of usable neutrons.

A further possibility to select the energy of a broadband neutron signal are monochromator crystals, for example pyrolytic graphite. The neutrons are scattered at the crystal lattice planes of such a crystal in accordance with the Bragg condition

$$n\,\lambda = 2\,d\sin\theta \quad ,$$

where n is a natural number and denotes the order of refraction, λ the neutron wavelength, d the distance of the lattice planes and θ the scattering angle with respect to the lattice plane normal. Accordingly, the energy of neutrons directed to the sample can be adjusted by selecting the orientation of the crystal with respect to the incident neutrons and the subsequent measurement setup.

2.5 Particle Transport and The Monte Carlo Method

For the simulation of particle transport problems different methods of calculations exist which can be divided into deterministic and stochastic methods. One of the former is the discrete ordinate method, in which the phase space of the problem is divided into a discrete mesh. The equation of state, which describes the dynamics of the system and generally represents a non-analytic (integro-) differential equation, is solved numerically for each node of the mesh, whereby the boundary conditions are determined by the (finite number of) adjacent nodes. In the case of particle transport problems, this is the *Boltzmann equation* in the six-dimensional phase space, spanned by the momentum-vector $\vec{p}$ and spatial coordinates $\vec{r}$ of the particles, which reads

$$\left(\frac{\partial}{\partial t} + \frac{1}{m}\,\vec{p} \cdot \vec{\nabla}_{\vec{r}} + \propto\vec{\nabla}_{\vec{p}}\right) f(\vec{r}, \vec{v}, t) = \left.\frac{\partial f}{\partial t}\right|_{\text{collision}} ,$$

where $f(\vec{r}, \vec{v}, t)$ describes the state of a particle, $\vec{F}$ external forces and $\frac{\partial f}{\partial t}|_{\text{collision}}$ the collision term describing the effect of collisions between the particles. The Boltzmann equation originates from the kinetic theory of gases and can also be applied to neutron transport problems [52]. In deterministic simulations, all information describing the phase space is supplied with the previously defined resolution.

In contrast, the state equations of a system are not solved directly by the Monte Carlo method. Instead, the state function is approximated by calculating many individual *histories* of particles and subsequent averaging. A large number of histories is often required in order to sample the relevant areas of the phase space with sufficient statistical precision. Thereby only some parameters of their average behavior (for example flux or derived quantities like dose rate) are recorded by the definition of tallies, which can be considered as two- or three-dimensional detector regions in the phase space [70].

The events of a history, for example, elastic scattering or fission, are stochastic processes whose input parameters (e.g., the branching ratio and the partial cross sections) and output parameters (energy and direction of the scattering products) are given by probability distribution functions. These distribution functions must be known for the simulation, typically in the form of nuclear databases. Using a pseudorandom number generator,[4] the parameters of each event are randomly set according to the underlying distribution functions. For a sufficient period length of the pseudorandom generator the resulting histories are unique, thus being called *random walks*, so that in principle the entire phase space of the problem can be sampled.

2.5.1 The Monte Carlo N-Particle Code MCNP

The Monte Carlo method is used in a variety of simulation tools to solve a wide range of different tasks. For the particle transport, especially for neutrons, the *Monte Carlo N-Particle Transport Code* MCNP is a powerful tool which can handle complex geometries and material compositions. Among the most important features are the

[4] A pseudorandom number generator is an algorithm for generating a sequence of numbers which are approximately randomly distributed. The sequence is not truly random, because it is completely determined by an initial value, called the seed.

extensive possibilities to define the source of starting particles, the calculation of criticality eigenvalues of systems with nuclear fuel, powerful geometry, tally and cross section plotter as well as advanced variance reduction techniques.

The development of MCNP began in 1957 at the Los Alamos National Laboratory and has been steadily extended. In 1994 the further development of MCNP was forked, which resulted in MCNP5 [71] and MCNPx [72]. MCNPx is based on version 4B (1994) and offers, as the first version, the possibility to transport any kind of particles over a broad energy range. Since nuclear data is unavailable for many particle types, especially heavy ions, those are treated with sophisticated physical models. The add-on MCUNED [73] offers the possibility to use databases for at least light ions (e.g., deuterons or alpha particles), see Sect. 2.5.4. With MCNP5, on the other hand, only classical particles (neutrons, electrons, photons) can be simulated. For these particles MCNP5 offers an advantage in computing speed compared to other MCNP versions. The most recent stable version is MCNP6.1 [74], which is a merge of MCNP5 and MCNPx and combines their advantages. For this work MCNPx, MCNP5 and MCNP6.1 are used depending on the problem.

2.5.2 *Fundamental Mathematical Principles*

As Monte Carlo results are obtained by averaging the contributions from many random walk histories, these results are subject to a statistical uncertainty which is a measure of the precision of the Monte Carlo simulation.

For a physical quantity such as the neutron flux at a specific (tally-) position to be measured in a simulation, the i-th random walk scores a contribution x_i to the tally. The expected value of x is then given by

$$E(x) = \int x\, f(x)\, \mathrm{d}x \quad ,$$

where $f(x)$ is the probability distribution that a given random walk provides the contribution x to the tally. In general, $f(x)$ is unknown but is implicitly determined by the Monte Carlo method. Thereby, the expected value $E(x)$ of the quantity is approximated by the sampled mean

$$\bar{x} = \frac{1}{N} \sum_{i=1}^{N} x_i \quad , \tag{2.10}$$

wherein x_i is the tally contribution from the i-th history and N the total number of histories. This statement is motivated by the strong law of large numbers [75] which implies that $\lim_{N\to\infty} \bar{x} = E(x)$ if $E(x)$ is finite and is thus a fundamental theorem of the Monte Carlo method.

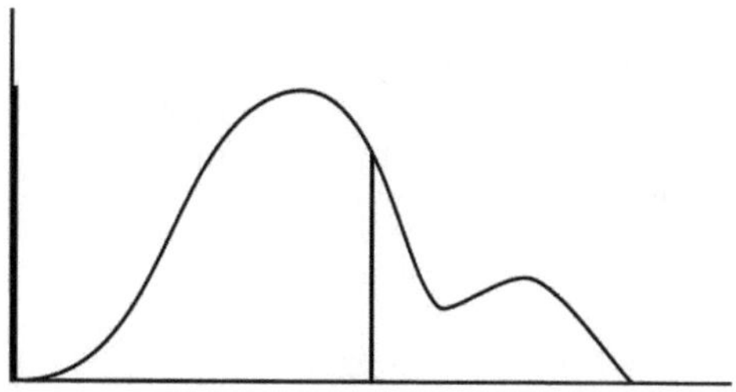

Fig. 2.9 Arbitrary distribution of individual scores x_i to a tally for a specific subset of histories. The thick black vertical line on the left indicates the zero scores to the tally. Courtesy of [77]

This also applies to the variance

$$\sigma^2 = \int [x - E(x)]^2 f(x)\,\mathrm{d}x = E(x^2) - E^2(x) \quad ,$$

of the calculated scores to the physical quantity which can be approximated by the Monte Carlo method as

$$S^2 = \frac{1}{N-1} \sum_{i=1}^{N} (x_i - \bar{x})^2 \tag{2.11}$$

for large N. Since S is the estimated standard deviation of all N sampled scores x_i, the estimated standard deviation of the estimated mean $\bar{x}$ is defined as

$$S_{\bar{x}} = \sqrt{\frac{S^2}{N}} \tag{2.12}$$

It is evident from Eq. 2.12 that the standard deviation $S_{\bar{x}} \propto \sqrt{N}^{-1}$ which implies that four times the computation time is needed to halve the standard deviation of the tally in question.

To assess the precision of Monte Carlo calculations it is necessary to define confidence intervals for the quantities x. As a second fundamental theorem of the Monte Carlo method, the central limit theorem applies which states that the distribution of mean values $\bar{x}$ of different sets of histories with the same size approaches a normal distribution where the mean of this distribution is given by $E(x)$ [76]. This holds for any distribution of scored values of x, as exemplarily depicted in Fig. 2.9.

In particular, the central limit theorem states for $\lim_{N\to\infty}$ that

$$P\left[\alpha\, S_{\bar{x}} < \sqrt{N}\,\frac{\bar{x} - E(x)}{\sigma} < \beta\, S_{\bar{x}}\right] = \frac{1}{\sqrt{2\,\pi}} \int_{\alpha}^{\beta} \exp\left(\frac{-t^2}{2}\right) \mathrm{d}t \quad ,$$

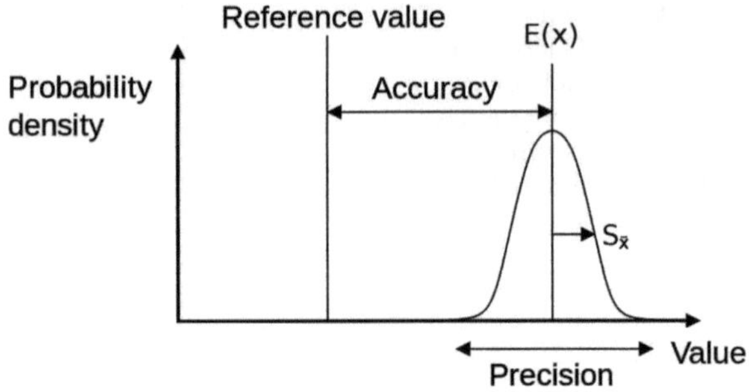

Fig. 2.10 Schematic demonstrating the difference between accuracy and precision of a Monte Carlo simulation. The mean value of the score distribution is denoted by $E(x)$ and the estimated standard deviation of the estimated mean $\bar{x}$ by $S_{\bar{x}}$. Courtesy of [78]

where the probability P of the estimated mean $\bar{x}$ lying in a confidence interval spanned by $\alpha\, S_{\bar{x}}$ and $\beta\, S_{\bar{x}}$ around the expected value value $E(x)$ amounts to the area of the normal distribution spanned by α and β.

Care should be taken when assessing the quality of a Monte Carlo result regarding the terms *accuracy* and *precision*. As indicated above the latter refers to the statistical uncertainty measured by the estimated standard deviation $S_{\bar{x}}$, whereas accuracy refers to how close the calculated mean value $E(x) \approx \bar{x}$ is to the true value of the physical quantity, see Fig. 2.10. The accuracy refers to the systematic error which is often elusive to the simulation.

2.5.3 Tallies, Error Estimation and Variance Reduction

As already stated in Sect. 2.5, in Monte Carlo calculations information about the simulated system is provided only if these are explicitly requested by the user. This is done in MCNP by means of tallies, which act as detectors and determine various physical quantities such as the neutron fluence. In fact all quantities of the form

$$\int \Phi(E)\, f(E)\, \mathrm{d}E$$

can be recorded, whereby $\Phi(E)$ denotes the energy-dependent fluence and $f(E)$ is an arbitrary response function, e.g., a flux-to-dose conversion function. Tallies can be defined on surfaces or cells (volumes). The definition of a three-dimensional mesh is also possible, whereby each voxel corresponds to a volume tally. Such mesh tallies are not bound to the geometry of the simulation model, but are superimposed. Each tally can be subdivided into bins, for example energy of number of collisions.

Particles simulated in MCNP do not always correspond to physical particles. By means of variance reduction methods, for example, an actually absorbed particle can be followed further, but its weight is reduced. A particle with weight 1 behaves analogously to a physical particle. In order to account for the law of particle conservation, the particles themselves are not tallied in MCNP, but rather their weights. In MCNP, the conservation of the number of particles is therefore reflected as a weight conservation law.

To each tally (bin) value the estimated relative error, cf. Sect. 2.5.2, is printed which is defined as

$$R \equiv \frac{S_{\bar{x}}}{\bar{x}} = \sqrt{\frac{\sum_{i=1}^{N} x_i^2}{(\sum_{i=1}^{N} x_i)^2} - \frac{1}{N}}$$

according to Eqs. 2.10, 2.11, and 2.12. It is evident that for identical and finite x_i $R = 0$. If all $x_i = 0$, then R is defined to be 0. The relative error can be used to estimate the quality of the calculated value of the tally bin of interest. Thus, tally values with $R < 0.1$ are considered reliable, except for point detector-tallies, for which $R < 0.05$ is required.

Variance reduction techniques can be applied to reduce the relative error of the desired tallies or to save computing time. The simplest methods are aimed at truncating areas of the phase space which do not contribute significantly to the desired result. This applies to the geometry as well as, for example, the energy.

Population control methods are used to increase the particle population in important areas of the phase space and to reduce it in unimportant areas. Important and unimportant areas are defined by the parameter *importance* in the simulation model. A sophisticated approach for this are *weight windows*, where depending on the energy, the position, or both, low-weighted particles play *Russian roulette* and high–weighted particles are split. Particles which lose Russian roulette are eliminated while the weight of particles which win Russian roulette are increased. If a high-weighted particle is split its products are reduced in weight. In average the weight of all particles as they contribute to any tally is conserved.

In certain cases, partly deterministic methods are also suitable which circumvent the random walk process at defined locations in the geometry and transport the particles deterministically, for example, point detector tallies. Apart from MCNP, there are also hybrid methods available which can generate optimized weight window distributions using deterministic and Monte Carlo methods, e.g., FW-CADIS [79].

2.5.4 Nuclear Data, Physical Models and Parallelization

As explained in Sect. 2.5, for an accurate solution of particle transport problems, databases containing the scattering cross sections of the incident particles for the isotopes of the model materials used as well as energy and angular distributions of the reaction products, are essential. For MCNP, extensive, pointwise, continuous-energy nuclear and atomic databases are provided. These involve the Evaluated Nuclear Data File (ENDF), the Advanced Computational Technology Initiative (ACTI), the Evaluated Nuclear Data Library (ENDL), the Evaluated Photon Data Library (EPDL), the Activation Library (ACTL), and compilations from Livermore and the Nuclear Physics Group at Los Alamos (T-16). The photon and electron data are mostly atomic rather than nuclear. Photonuclear data is also included which, for example, allow for the treatment of neutron production from high-energy bremsstrahlung-photons.

There are many other working groups that also provide nuclear databases for general and special purposes. These include, for example, the Japanese Nuclear Data Library (JENDL), the Joint Evaluated Fission and Fusion File (JEFF), the Russian File for Evaluated Nuclear Data (RUSFOND), the Russian Evaluated Neutron Data Library (BROND), the Chinese Evaluated Nuclear Data Library (CENDL) as well as the TALYS Evaluated Nuclear Data Library (TENDL) based on the TALYS code system.

To ensure a high quality, experimental nuclear data results are evaluated by a Nuclear Data Organization. Nuclear models are used for nuclides or energy ranges for which only few measured data is available, the parameters of which are benchmarked and adjusted with these measurement data. The resulting data library contains a combination of interpolatable data points and resonance parameters. Evaluated data can be processed into the pointwise ACE (*A Compact ENDF*) format, which is demanded by MCNP, by codes such as NJOY [49].

The materials are treated in MCNP as a nuclide mixture of a free gas. As discussed in Sect. 2.3.2 additional effects to the particle-particle scattering have to be taken into consideration for solids or fluids if the neutron's de Broglie wavelength is in the range of atomic bonds in the molecules or crystal lattices. These effects are accounted for by modifications of the cross section according to a scattering function $S(\alpha, \beta)$ as given in Eq. 2.7. Data tables for $S(\alpha, \beta)$ scattering treatment is available in MCNP for a limited number of substances and temperatures mostly relevant to thermal reactors and cold moderators.

For nuclides or energy ranges for which no databases are available, MCNP offers built-in physics models such as the Cascade Exciton Model (CEM), the intranuclear cascade models Bertini, ISABEL, INCL with the abrasion ablation model ABLA, and the Quark-Gluon String Model (LAQGSM).

MCNP can be executed very efficiently on multi-processor systems, for example HPC clusters, since the particles are successively sampled in an internal program loop. In this case, MCNP can be used with the two parallelization paradigms OMP (open multi-processing) and MPI (message passing interface). When using OMP, only one instance of MCNP is executed, with the internal particle loop of MCNP being split onto several processor cores. These threads share the memory allocated to MCNP, which is why OMP runs are also referred to as shared memory multi-processing. Due to the memory being shared the performance suffers when several physical CPUs are requested. However, OMP can only be used for classical particles (neutrons, photons, electrons) and tabulated cross sections, not for physical models.

In MPI runs, several individual instances of MCNP are executed in parallel. Since each instance has its own memory allocated, there is a master process distributing cross section data to the slave processes and setting rendezvous points during the simulation where the statistics of the tallies are accumulated. Because of the large overhead that arise when using many processors, MPI does not scale linearly and the performance will suffer.

If calculations are to be performed on many CPU cores, it is useful to combine OMP and MPI in *hybrid jobs*. A predefined number of OMP threads is executed for each MPI slave process. Since each MPI process allocates its own memory, which is then shared by the corresponding OMP threads, the sweet spot is to start one MPI process per physical CPU, which then runs one OMP thread on each core of the CPU.

References

1. J.D. Bjorken, S.D. Drell, *Relativistic Quantum Mechanics*. (MC Graw Hill Book Company, 1964). ISBN: 0-07-005493-2
2. G. Musiol, J. Ranft, R. Reif, D. Seliger, *Kern- und Elementarteilchenphysik*, 2nd edn. (Verlag Harri Deutsch, 1995), p. 1127. ISBN: 3-8171-1404-4
3. H. Schwoerer, J. Magill, B. Beleites, Lasers and nuclei. Applications of ultrahigh intensity lasers in nuclear science. Lecture Notes in Physics, vol. 694 (Springer, Berlin, 2006). ISBN: 978-3-540-30271-1. https://doi.org/10.1007/b11559214
4. G. Schmidt, *Physics of High Temperature Plasmas*, 2nd edn. (Academic Press, New York, 1979)
5. R.A. Snavely et al., Intense high-energy proton beams from petawatt-laser irradiation of solids. Phys. Rev. Lett. **85**(14), 2945–2948 (2000). https://doi.org/10.1103/PhysRevLett.85.2945
6. M. Hegelich et al., MeV ion jets from short-pulse-laser interaction with thin foils. Phys. Rev. Lett. **89**(8), Aug 2002. https://doi.org/10.1103/PhysRevLett.89.085002
7. Stephen P. Hatchett et al., Electron, photon, and ion beams from the relativistic interaction of Petawatt laser pulses with solid targets. Phys. Plasmas **7**(5), 2076–2082 (2000)
8. J. Fuchs et al., Laser-driven proton scaling laws and new paths towards energy increase. Nat. Phys. **2**(1), 48–54 (2006). https://doi.org/10.1038/nphys199
9. A.P.L. Robinson, A.R. Bell, R.J. Kingham, Effect of target composition on proton energy spectra in ultraintense laser-solid interactions. Phys. Rev. Lett. **96**(3) (2006). https://doi.org/10.1103/PhysRevLett.96.035005
10. S. Karsch, High-intensity laser generated neutrons: a novel neutron source and new tool for plasma diagnostics. Ph.D. Thesis. München: LMU Munich, Faculty of Physics (2002). https://edoc.ub.uni-muenchen.de/703/1/Karsch_Stefan.pdf
11. H. Schwoerer et al., Laser-plasma acceleration of quasi-monoenergetic protons from microstructured targets. Nature **439**(7075), 445–448 (2006). https://doi.org/10.1038/nature04492
12. L. Yin et al., Monoenergetic and GeV ion acceleration from the laser breakout afterburner using ultrathin targets a. Phys. Plasmas **14**, 5 (2007)
13. B.J. Albright et al., Relativistic buneman instability in the laser breakout afterburner. Phys. Plasmas **14**, 9 (2007)
14. A. Henig et al., Enhanced laser-driven ion acceleration in the relativistic transparency regime. Phys. Rev. Lett. **103**(4) (2009). https://doi.org/10.1103/PhysRevLett.103.045002
15. D. Jung, Ion acceleration from relativistic laser nano-target interaction. Ph.D. Thesis. München: LMU Munich, Faculty of Physics (2012). https://edoc.ub.uni-muenchen.de/14074/1/Jung_Daniel.pdf
16. M. Roth, Breaking the 70 MeV proton energy threshold in laser proton acceleration and guiding beams to applications, in *IPAC Dresden* (2014). http://accelconf.web.cern.ch/accelconf/IPAC2014/talks/wexb01_talk.pdf
17. D. Jung et al., Characterization of a novel, short pulse laser-driven neutron source. Phys. Plasmas **20**(5), 056706 (2013). p. 9. https://doi.org/10.1063/1.4804640

18. M. Roth et al., Bright laser-driven neutron source based on the relativistic transparency of solids. Phys. Rev. Lett. **110**(4) (2013). https://doi.org/10.1103/PhysRevLett.110.044802
19. D. Faircloth, Ion sources for high-power hadron accelerators, in *CERN Accelerator School on High Power Hadron Machines (CAS 2011)*. (Bilbao, Spain, May 24–June 2, 2011) (2013). arXiv:1302.3745v1 [physics.acc-ph]. https://inspirehep.net/record/1220608/files/arXiv:1302.3745.pdf
20. R. Scrivens, Electron and ion sources for particle accelerators, in *CAS - CERN Accelerator School: Intermediate Course on Accelerator Physics*. (Zeuthen, Germany, Sept, 15–26, 2003) (2006), pp. 495–504. https://doi.org/10.5170/CERN-2006-002.495. https://cds.cern.ch/record/941321
21. I.M. Kapchinskii, V.A. Teplyakov, Linear Ion Accelerator with spatially homogeneous strong focusing. Instrum. Exp. Tech. **2.322**(6) (1970)
22. C. Zhang, A. Schempp, Beam dynamics studies on a 200 mA proton radio frequency quadrupole accelerator. Nucl. Instrum. Methods Phys. Res. Sect. A Accel. Spectrometers Detect. Assoc. Equip. **586**(2), 153–159 (2008). ISSN: 0168-9002. https://doi.org/10.1016/j.nima.2007.12.001. http://www.sciencedirect.com/science/article/pii/S0168900207024199
23. R.H. Stokes, T.P. Wangler, Radiofrequency quadrupole accelerators and their applications. Annu. Rev. Nucl. Part. Sci. **38**(1), 97–118 (1988). https://doi.org/10.1146/annurev.ns.38.120188.000525
24. F. Hinterberger, *Physik der Teilchenbeschleuniger und Ionenoptik*, 2nd edn. (Springer, 2008)
25. Superconductivity and cryogenics for accelerators and detectors, in *CAS - CERN Accelerator School*. (Erice, Italy, May 8–17, 2002) (2004). ISBN: 92-9083-230-4. http://cds.cern.ch/record/503603/files/CERN-2004-008.pdf?version=2
26. L. Lilje et al., Achievement of 35 MV/m in the superconducting nine-cell cavities for TESLA. Nucl. Instrum. Methods Phys. Res. Sect. A Accel. Spectrometers Detect. Assoc. Equip. **524**(1–3), 1–12 (2004). ISSN: 0168-9002. https://doi.org/10.1016/j.nima.2004.01.045. http://www.sciencedirect.com/science/article/pii/S0168900204001391
27. G.D. Alton, J.R. Beene, The holifield radioactive ion beam facility at the oak ridge national laboratory: present status and future plans. J. Phys. G Nucl. Part. Phys. **24**(8), 1347 (1998). http://iopscience.iop.org/article/10.1088/0954-3899/24/8/006/pdf
28. High Voltage Engineering Europa B.V. Singletron Accelerator Systems. Coaxial and In-Line Positive Ion Accelerators. Online brochure. Netherlands. http://www.highvolteng.com/media/Brochures/Singletron_broch.pdf (visited on 17 May 2017)
29. K. Bethge, G. Walter, B. Wiedemann. *Kernphysik*. 2nd edn. (Springer, 2001). 196 ff. ISBN: 978-3-540-74567-9
30. J. Lilley. *Nuclear Physics. Principles and Applications*. Repr. with Corrections 2006. (Wiley, 2001). ISBN: 978-0-471-97936-4
31. A.J. Koning, D. Rochman, Modern nuclear data evaluation with the TALYS code system. Nucl. Data Sheets **113**(12), 2841–2934 (2012). ISSN: 0090-3752. https://doi.org/10.1016/j.nds.2012.11.002. http://www.sciencedirect.com/science/article/pii/S0090375212000889
32. Passage of particles through matter, in: *2016 Review of Particle Physics*. Ed. by C. Patrignani et al. Revised Jan. 2012 by H. Bichsel, D.E. Groom, S.R. Klein. June 18, 2012, pp. 18–20. http://pdg.lbl.gov/2013/reviews/rpp2012-rev-passage-particles-matter.pdf (visited on 28 Apr 2017)
33. H.A. Kramers. XCIII. On the theory of X-ray absorption and of the continuous X-ray spectrum. In: Philosophical Magazine Series 6 46.275, 1923, pp. 836–871. https://doi.org/10.1080/14786442308565244
34. G.C. Baldwin, G.S. Klaiber, Photo-fission in heavy elements. Phys. Rev. **71**(1), 3–10 (1947). https://doi.org/10.1103/PhysRev.71.3
35. C. Segebade, H.P. Weise, G.J. Lutz, *Photon Activation Analysis*. Berlin, Germany: de Gruyter (1988). ISBN 3-11-007250-5
36. Y. Otake, Introduction of RIKEN miniature neutron source system “RANS”. Technical Report Japan (2013). http://rans.riken.jp/pdf/14012401.pdf

37. Y. Yamagata, J. Ju, K. Hirota, Neutron generation source, and neutron generation device. EP Patent App. EP20130757266 (Japan) (2015). http://www.google.com/patents/EP2824999A1?cl=en
38. H. Kumada et al., Development of beryllium-based neutron target system with three-layer structure for accelerator-based neutron source for boron neutron capture therapy. Appl. Radiat. Isot. **106**, 78–83 (2015). The 16th International Congress on Neutron Capture Therapy (ICNCT-16). Representative person of the Organizing Committee: Dr Hanna Koivunoro (Secretary general of the ICNCT-16). ISSN: 0969-8043. https://doi.org/10.1016/j.apradiso.2015.07.033. http://www.sciencedirect.com/science/article/pii/S0969804315301159
39. T. Rinckel, D.V. Baxter, J. Doskow, P.E. Sokol, T. Todd, Target performance at the low energy neutron source. Phys. Procedia **26**, 168–177 (2012). ISSN: 1875-3892. https://doi.org/10.1016/j.phpro.2012.03.022. http://www.sciencedirect.com/science/article/pii/S1875389212004385
40. H. Kumada et al., Development of beryllium-based neutron target system with three-layer structure for accelerator-based neutron source for boron neutron capture therapy. Appl. Radiat. Isot. **106**, 78–83 (2015). The 16th International Congress on Neutron Capture Therapy (ICNCT-16). Representative person of the Organizing Committee: Dr Hanna Koivunoro (Secretary general of the ICNCT-16). ISSN: 0969-8043. https://doi.org/10.1016/j.apradiso.2015.07.033. http://www.sciencedirect.com/science/article/pii/S0969804315301159
41. P. Zakalek, Development of high-brilliant neutron source targets. Workshop Presentation in Unkel, Germany. Jülich Center for Neutron Science 2, Forschungszentrum Jülich (2016)
42. H. Wipf, Solubility and diffusion of hydrogen in pure metals and alloys. Phys. Scr. 2001.T94, 43 (2001). http://stacks.iop.org/1402-4896/2001/i=T94/a=006
43. G. Alefeld, J. Völkl (eds.), Hydrogen in Metals I. Basic Properties. Vol. 28. Topics in Applied Physics (Springer, Heidelberg, 1978). ISBN: 978-3- 540-08705-2. https://doi.org/10.1007/3-540-08705-2
44. E. Abramov, M.P. Riehm, D.A. Thompson, W.W. Smeltzer, Deuterium permeation and diffusion in high-purity beryllium. J. Nucl. Mater. **175**(1), 90–95 (1990). ISSN: 0022-3115. https://doi.org/10.1016/0022-3115(90)90274-Q
45. P. Zakalek et al., Temperature profiles inside a target irradiated with protons or deuterons for the development of a compact accelerator driven neutron source. Phys. B Condens. Matter **ISSN**, 0921–4526 (2018). https://doi.org/10.1016/j.physb.2018.01.049
46. S. Gary, *Was* (Springer, Fundamentals of Radiation Materials Science. Metals and Alloys, 2007). ISBN: 978-3-540-49471-3. https://doi.org/10.1007/978-3-540-49472-0
47. Y. Dai et al., Assessment of the lifetime of the beam window of MEGAPIE target liquid metal container. J. Nucl. Mater. **356**(1), 308–320 (2006). Proceedings of the Seventh International Workshop on Spallation Materials Technology. ISSN: 0022-3115. https://doi.org/10.1016/j.jnucmat.2006.05.036
48. M.J. Norgett, M.T. Robinson, I.M. Torrens, A proposed method of calculating displacement dose rates. Nucl. Eng. Des. **33**(1), 50–54 (1975). ISSN: 0029-5493. https://doi.org/10.1016/0029-5493(75)90035-7
49. R.E. MacFarlane, The NJOY Nuclear Data Processing System, Version 2012. Ed. by A. C. Kahler. LA-UR-12-27079. Technical Report Los Alamos National Laboratory (2012). http://t2.lanl.gov/nis/publications/NJOY2012.pdf
50. B.L. Cohen, T.H. Handley, experimental studies of (p, t) reactions. Phys. Rev. **93**(3), 514–517 (1954). https://doi.org/10.1103/PhysRev.93.514
51. F.J. Bermejo, F. Sordo, ESS-Bilbao Target station. Technical Design Report. ESS-Bilbao Target Devision, p. 33 (2013). ISRN: 978-84-695-8105-6. http://www.essbilbao.org/index.php/en/component/docman/doc_download/417-essb-ntar-2013-final-tdr-july
52. D. Emendörfer, K.H. Höcker, *Theorie der Kernreaktoren*, vol. 2, Stuttgart, Germany (1969). ISBN: 3-411-01599-3
53. D.E. Parks, M.S. Nelkin, J.R. Beyster, N.F. Wikner. *SlowNeutron Scattering and Thermalization* (W. A. Benjamin Inc., New York, 1970)
54. R.E. MacFarlane, New Thermal Neutron Scattering Files for ENDF/B-VI Release 2. LA-12639-MS. Los Alamos National Laboratory, USA, Mar 1994. https://t2.lanl.gov/nis/publications/thermal.pdf (visited on 06/21/2017)

55. THERMR. Incoherent Inelastic Scattering. Los Alamos National Laboratory (2013). http://t2.lanl.gov/nis/njoy/ther02.html (visited on 05/25/2017)
56. C.M. Lavelle et al., Neutronic design and measured performance of the Low Energy Neutron Source (LENS) target moderator reflector assembly. Nucl. Instrum. Methods Phys. Res. Sect. A Accel. Spectrometers Detect. Assoc. Equip. **587**(2–3), 324–341 (2008). ISSN: 0168-9002. https://doi.org/10.1016/j.nima.2007.12.044. http://www.sciencedirect.com/science/article/pii/S0168900208000028
57. T.A. Broome, Prospects for targets and methane moderators at ISIS, in *Proceedings of the Meetings ICANS-XIII and ESS-PM4 2* (1995). ISSN: 1019-6447. https://inis.iaea.org/search/searchsinglerecord.aspx?RN=27061066
58. Q.X. Feng et al., The solution of cold neutron source using solid methane moderator for the CPHS. Phys. Procedia **26**, 49–54 (2012). ISSN: 1875-3892. https://doi.org/10.1016/j.phpro.2012.03.008. http://www.sciencedirect.com/science/article/pii/S1875389212004245
59. J.M. Carpenter, Thermally activated release of stored chemical energy in cryogenic media. Nature **330**, 358–360 (1987). ISSN: 0028-0836. https://doi.org/10.1038/330358a0
60. E. Shabalin et al., Experimental study of swelling of irradiated solid methane during annealing. Nucl. Instrum. Methods Phys. Res. Sect. B Beam Interact. Mater. Atoms **266**(24), 5126–5131 (2008). ISSN: 0168-583X. https://doi.org/10.1016/j.nimb.2008.09.022. http://www.sciencedirect.com/science/article/pii/S0168583X08010379
61. M. Huerta Parajon, E. Abad, F.J. Bermejo, A review of the cold neutron moderator materials: neutronic performance and radiation effects. Phys. Procedia **60**, 74–82 (2014). ISSN: 1875-3892. https://doi.org/10.1016/j.phpro.2014.11.012. http://www.sciencedirect.com/science/article/pii/S1875389214005598
62. V. Ananiev et al., Pelletized cold moderator of the IBR-2 reactor: current status and future development. J. Phys. Conf. Ser. **746**(1) (2016). https://doi.org/10.1088/1742-6596/746/1/012031
63. J. Van Kranendonk, Solid Hydrogen. Theory of the Properties of Solid H_2, HD, and D_2. Springer US (1983). ISBN: 978-1-4684-4303-5. https://doi.org/10.1007/978-1-4684-4301-1
64. K.B. Grammer et al., Measurement of the scattering cross section of slow neutrons on liquid parahydrogen from neutron transmission. Phys. Rev. B **91**(18) (2015). https://doi.org/10.1103/PhysRevB.91.180301
65. K. Batkov, A. Takibayev, L. Zanini, F. Mezei, Unperturbed moderator brightness in pulsed neutron sources. Nucl. Instrum. Methods A Accel. Spectrometers Detect. Assoc. Equip. **729**, 500–505 (2013). ISSN: 0168- 9002. https://doi.org/10.1016/j.nima.2013.07.031. http://www.sciencedirect.com/science/article/pii/S0168900213010073
66. Ferenc Mezei et al., Low dimensional neutron moderators for enhanced source brightness. J. Neutron Res. **17**(2), 101–105 (2014). https://doi.org/10.3233/JNR-140013. https://arxiv.org/pdf/1311.2474
67. V.F. Sears. *Neutron Optics. An introduction to the theory of neutron optical phenomena and their applications* (Oxford University Press, 1989). ISBN: 978-0195046014
68. T. Brückel, D. Richter, G. Roth, A. Wischnewski, R. Zorn (eds.), Neutron Scattering: Lectures, vol. 106. Schlüsseltechnologien/Key Technologies. Forschungszentrum Jülich GmbH (2015). ISBN 978-3-95806-055-5
69. T. Chupp, Neutron Optics and Polarization. Presentation at the Summerschool on Neutron Research. NIST Center for Neutron Research (2009). https://www.ncnr.nist.gov/summerschool/ss09/pdf/Chupp_FP09.pdf (visited on 06/15/2017)
70. N. Metropolis, S. Ulam, The monte carlo method. J. Am. Stat. Assoc. **44**(247), 335–341 (1949). PMID: 18139350. https://doi.org/10.1080/01621459.1949.10483310
71. X-5 Monte Carlo Team. MCNP - A General N-Particle Transport Code, Version 5. LA-UR-03-1987. Technical Report Los Alamos National Laboratory, 1 Feb 2008. https://laws.lanl.gov/vhosts/mcnp.lanl.gov/pdf_files/la-ur-03-1987.pdf
72. D.B. Pelowitz (ed.), MCNPX Users Manual Version 2.7.0. LA-CP-11-00438. Los Alamos, USA (2011)

73. P. Sauvan, J. Sanz, F. Ogando, New capabilities for Monte Carlo simulation of deuteron transport and secondary products generation. Nucl. Instrum. Methods Phys. Res. (Section A) **614**(3), 323–330 (2010). ISSN: 0168-9002. https://doi.org/10.1016/j.nima.2009.12.084. http://www.sciencedirect.com/science/article/pii/S0168900209024255/pdfft?md5=2e82c9c5e77cab8bf504087eaf062bbe&pid=1-s2.0-S0168900209024255-main.pdf
74. T. Goorley et al., Initial MCNP6 release overview. Nucl. Technol. **180**(3), 298–315 (2012). https://doi.org/10.13182/NT11-135
75. M. Loeve, Probability theory I, in *Graduate Texts in Mathematics*, 4th edn., vol. 45 (Springer New York, 1977). ISBN: 978-0-387-90210-4. https://doi.org/10.1007/978-1-4684-9464-8
76. J. Rice. *Mathematical Statistics and Data Analysis*, 2nd edn. (Duxbury Press, 1995). ISBN: 0-534-20934-3
77. Optimsmoose. File:Some probability distribution.png. The copyright holder of this work releases this work into the public domain. This applies worldwide. 5 Jan 2011. https://upload.wikimedia.org/wimimedia/commons/e/e5/Some_probability_distribution.png (visited on 24 May 2017)
78. Pekaje. File: Accuracy and precision.svg. Permission is granted to copy, distribute and/or modify this document under the terms of the GNU Free Documentation License, Version 1.2 or any later version published by the Free Software Foundation (2007). https://upload.wikimedia.org/wikipedia/commons/3/38/Accuracy_and_precision.svg (visited on 24 May 2017)
79. J.C. Wagner, D.E. Peplow, S.W. Mosher, FW-CADIS method for global and regional variance reduction of monte carlo radiation transport calculations. Nucl. Sci. Eng. **176**(1), 37–57 (2014). https://doi.org/10.13182/NSE12-33

Chapter 3
Optimization Studies on a Laser-Driven Neutron Source

As outlined in Sect. 2.1.1 only the most powerful lasers which are capable of producing ultra-short pulses are suitable for the acceleration of light ions. In this regime the efficient acceleration mechanisms are the Target Normal Sheath Acceleration (TSNA) and the Break-Out Afterburner (BOA), respectively, depending on the target thickness.

One of the most promising experiments on neutron production using laser-based ion acceleration recently took place at Trident Laser in Los Alamos. The results of the experiments are therefore used as input parameters for the investigation of the behavior of the moderator system.

First, the experimental setup is explained including the results from which a source definition for MCNP is derived. This source definition is used to carry out parameter studies for the design and optimization of the moderator system. In addition to the general structure of the moderator with regard to material, geometry, and cold neutron production, the idea of the Flux Finger Moderator is presented, whose innovative design significantly increases the brilliance of thermal neutrons. Finally, the impact of an additional reflector on the various performance parameters of the moderator system (e.g., brilliance and pulse structure) is investigated.

In all simulations discussed in this chapter the symmetry axis of the moderator is assumed to be in positive z-direction. Neutrons with an energy $E < 7\,\text{meV} \mathrel{\widehat{=}} 81\,\text{K}$ are considered as cold, whereas neutrons below $30\,\text{meV} \mathrel{\widehat{=}} 350\,\text{K}$ are considered as thermal, cf. Table A.1 in the appendix.

3.1 The Trident Experiment

A full characterization of a short-pulsed laser-driven neutron source was performed by [1–3] in 2012 at the Trident laser facility of the Los Alamos National Laboratory. As outlined in Sect. 2.1.1 the Trident laser is capable of accelerating ions in the BOA

J. P. Dabruck, *Target Station Optimization for the High-Brilliance Neutron Source HBS*, Springer Theses, https://doi.org/10.1007/978-3-030-05639-1_3

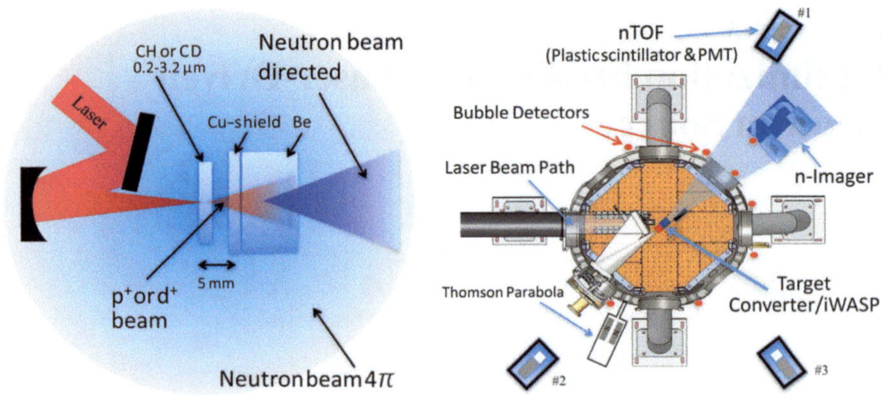

(a) Inside view of the target chamber. **(b)** Full setup with target chamber and detectors.

Fig. 3.1 Schematics of the experimental setup at Trident [1]

regime with appropriate targets, however, the TNSA regime was also investigated using a thicker target.

The setup is schematically shown in Fig. 3.1a. The laser beam is focused on the target foil using a parabolic mirror. On average 80 J were deposited on a focus point with a diameter of about 6 μm and a pulse duration of 600 fs. The peak intensity was approximately $5 \times 10^{20}\,\mathrm{W/cm^2}$. The target foil consisted of a thin layer of polyethylene (CH_2) or deuterated polyethylene (CD_2) with varying thickness between 0.1 μm and 3.2 μm. Neutrons are produced in a beryllium converter caused by the ^{9}Be(p,n) and ^{9}Be(d,n) reactions, respectively, as explained in Sect. 2.2.1.

The full setup of the experiments which were carried out in an evacuated target chamber, is shown in Fig. 3.1b. For diagnostics of the ion beam, the beryllium converter was replaced by an *ion wide angle spectrometer* (iWASP), whose entrance slit was additionally surrounded by several copper films in order to perform a *nuclear activation-based imaging spectroscopy* (NAIS). The iWASP is used to record the ion spectra for different angles whereas the NAIS films are used to determine the shape of the beam.

To measure neutron yield, spectrum and angular distribution commercial bubble detectors from Bubble Technology Industries (BTI), nTOF detectors (10 cm in diameter, 1.88 cm thick NE102 plastic scintillators coupled with fast 12.5 cm photo multiplier tubes), and a neutron imaging station were used.

The highest neutron yield was obtained with a 480 nm thick CD_2 target. Therefore, the experimental results are explained exemplarily for exactly this configuration. Results of the deuteron beam diagnostics are shown in Fig. 3.2. Interestingly, the deuteron spectrum measured with the iWASP is only slightly depending on the angle as indicated by Fig. 3.2a. A typical deuteron spectrum is shown in Fig. 3.2b for angles between 6° and 8°. On the basis of the iWASP measurements, the total number of

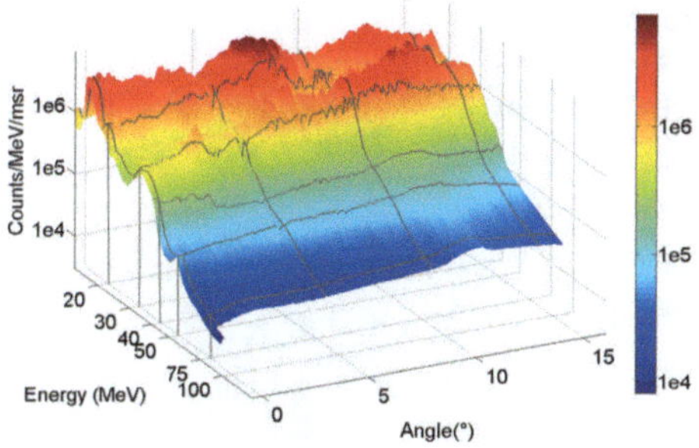

(a) Angularly resolved.

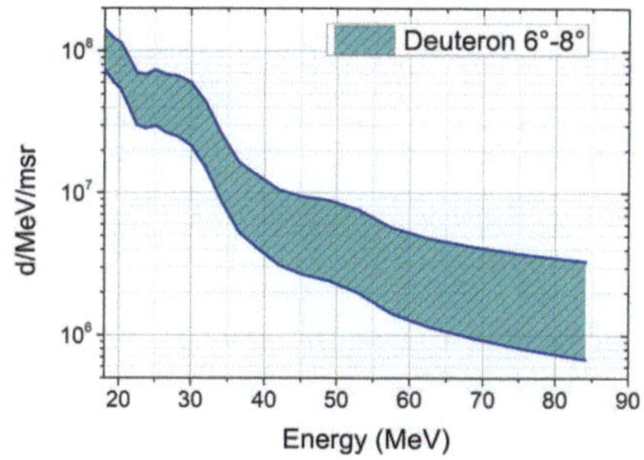

(b) Spectrum between 6° (upper line) and 8° (lower line)

Fig. 3.2 Spectra of deuterons produced with 480 nm thick CD_2 film [1]

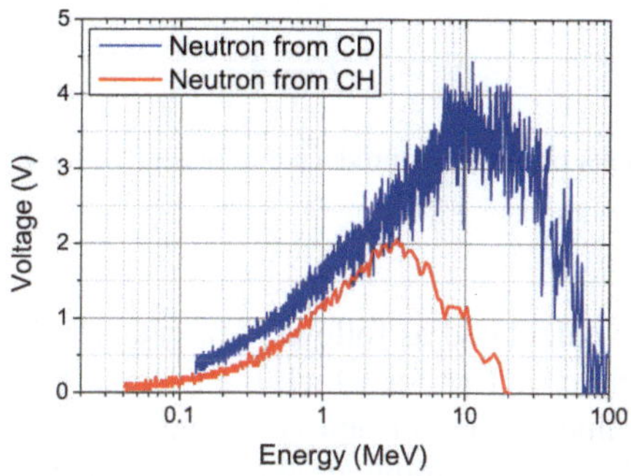

(a) Spectra of neutrons produced by protons (red) and deuterons (blue).

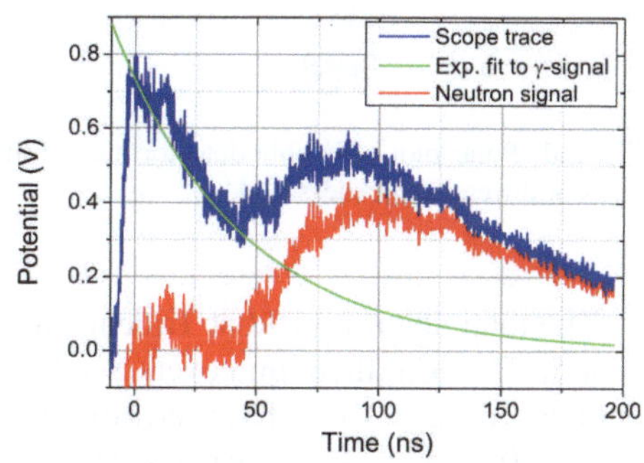

(b) Pulseshape for neutrons produced by deuterons. The γ-flash (green) is subtracted from the raw scope trace (blue) resulting in the neutron signal (red).

Fig. 3.3 Trident nTOF measurement results for neutrons produced by protons and deuterons in the BOA regime with a 400 nm thick CH_2 film and a 480 nm thick CD_2 film, respectively [1]

deuterons produced with energies $E > 15$ MeV (the lower threshold of the iWASP) could be estimated to $(5 \pm 3) \times 10^{11}$ per shot.

The spectrum of the generated neutrons shows a peak at approximately 18 MeV and extends up to 100 MeV according to the blue curve in Fig. 3.3a. The red curve corresponds to the neutron spectrum originating from CH_2 foils. The pulse structure of the neutrons is given by the red curve in Fig. 3.3b. Accordingly, the main pulse has a length of more than 200 ns and occurs after a 40 ns long prepulse.

The results of the measurements with the bubble detectors for thick and thin target foils are shown in the polar plot Fig. 3.4. In the TNSA regime (left diagram), a isotropic distribution of the neutrons occurs, which is also significantly lower than that of the neutrons produced by deuterons in the BOA regime (right diagram). In the latter case, a forward-directed component appears, which is significantly higher with 4.4×10^9 n/sr than the isotropic one with 2×10^9 n/sr.

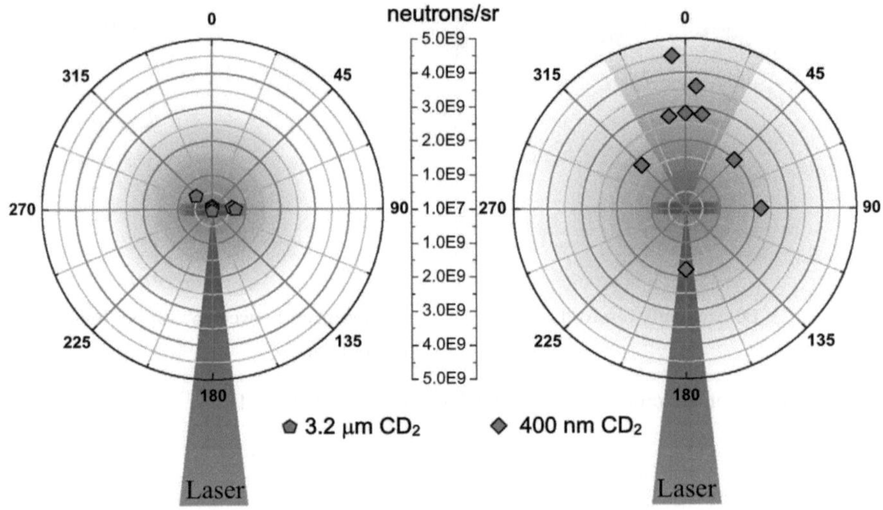

Fig. 3.4 Polar plot of bubble detector measurements on neutrons produced at Trident with CD_2 films of different thicknesses [1]

The total neutron yield at Trident can be calculated as follows: The blue shaded cone in forward direction covers a solid angle of 1 sr for which the average yield amounts to 3.2×10^9 n. The average of the values of the isotropic component amounts to 2×10^9 n/sr $\times (4\pi - 1)$sr $= 2.31 \times 10^{10}$n. By assuming an operating frequency of 10 Hz, which is a reasonable assumption for a future laser-driven system, a total neutron yield of 2.63×10^{11} n/s is obtained. This value is therefore used as the source strength of neutrons produced in the converter for all further simulations. This estimation can easily be verified by integrating over the linearly interpolated values of the angular distribution, cf. Fig. 3.8b in Sect. 3.2.3. This results in a value of 2.62×10^{11} n/s at 10 Hz, which is very close to the value from the simple estimate above.

3.2 Simulation Setup and Preparatory Work

Preliminary simulations are conducted to make the first material selection regarding the moderation performance. In addition the question of whether there are any harmful mechanisms in the regime that prevails at the Trident Laser is answered. For all further simulation studies a detailed source definition based on the measurement results at Trident is derived. The coordinate system is selected in such a way that the incident laser beam and thus also the preferred direction of the accelerated ions point along the positive z-axis being referred to as *forward direction*.

3.2.1 Moderation of a High-Energy Neutron Pencil Beam

Spectra of Neutrons behind Thermal Moderators In preliminary investigations, moderator blocks with a front surface of 40 cm × 40 cm and varying thickness are impinged by a neutrons pencil beam to assess the moderation properties of heavy water, beryllium, and graphite. Light water is neglected at this point, due to its high absorption cross section. The neutron energy is set to $E = 18$ MeV because the peak of the neutron spectrum generated with deuterons via BOA is located between 10 and 20 MeV according to Fig. 3.3a. The neutrons leaving the moderator are tallied directly on its rear surface.

The theoretical predictions shown in Sect. 2.3.1 for the moderation performance of common materials are confirmed by the spectra given in Fig. 3.5. Thicknesses of 10 cm and 50 cm are considered in the plots. Only beryllium can produce a reasonable thermal flux even at a small layer thickness of 10 cm. The flux maximum in the thermal range is higher by a factor of 8 than that of D_2O and by 4 orders of magnitude higher than that of graphite. Beryllium shows the highest thermal flux due to the very good slowing-down power and its high (n, 2n) cross section. The thermal flux is twice as high as for D_2O and a factor 30 higher than for graphite. The optimum thickness regarding the highest yield of neutrons in the energy range below 10^{-7} MeV amounts to 30 cm for beryllium, 45 cm for graphite and 35 cm for heavy water.

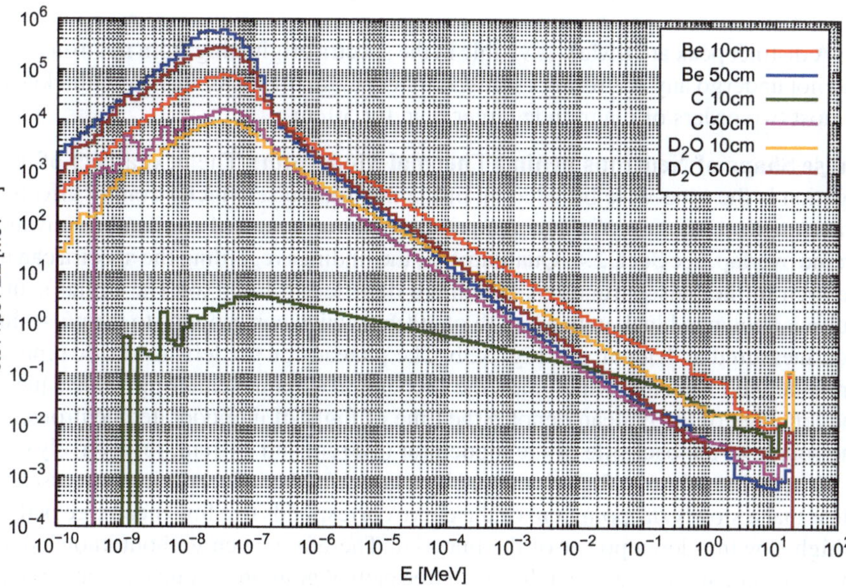

Fig. 3.5 Spectrum of neutrons with an initial energy of 18 MeV behind beryllium, graphite and heavy water layers with thicknesses of 10 and 50 cm, respectively

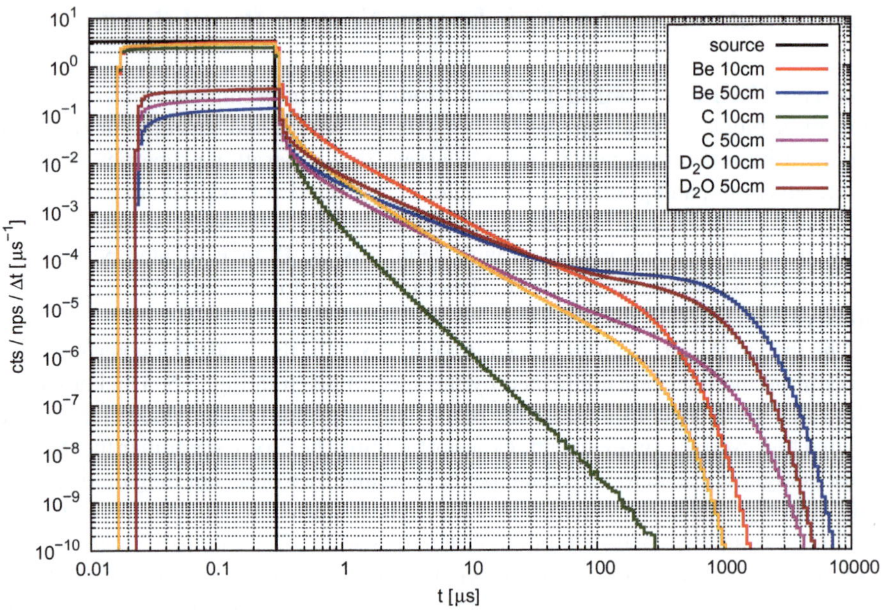

Fig. 3.6 Pulse shape of neutrons with an initial energy of 18 MeV behind beryllium, graphite and heavy water layers with thicknesses of 10 and 50 cm. The initial neutron beam has a pulse length of 300 ns which is the total pulse length of the neutrons produced by deuterons at Trident

A distinct peak at 18 MeV is visible in the spectra originating from neutrons which did not undergo any scattering events. For beryllium, the peak for 10 cm thickness is almost two orders of magnitude larger than for 50 cm.

Pulse Shape of Neutrons behind Thermal Moderators The next aspect of investigation is the time-wise broadening of the neutron pulse which strongly affects the wavelength resolution of the corresponding instruments and can lead to unwanted frame overlap, see Sect. 2.4. Qualitatively, the findings regarding the spectra can also be seen in the pulse structures, which are depicted in Fig. 3.6. The initial neutron pulse is shown in black with a length of 300 ns, which corresponds to the pulse duration measured at Trident, see Fig. 3.3b. The plateaus on the left-hand side represent unmoderated neutrons which pass the moderator without scattering. Consequently, the left edge of the plateau denotes the time of flight of unmoderated neutrons from the source to the backside of the moderator (100 cm and 140 cm, respectively).

Since a long time of flight is caused by a large number of scattering processes, it is clear that a slowly dropping tail of the curve corresponds to a low mean free path and a high slowing-down power of the material. Therefore, even without knowledge of the spectrum, it is evident that the mean free path of graphite is much larger compared to beryllium and heavy water. Due to its weak slowing-down power, the neutrons in graphite are heavily undermoderated resulting in a pulse with 280 μs. With heavy water and beryllium, pulse lengths of 1 ms or 1.5 ms, respectively, are reached, with the pulse height for beryllium being almost one order of magnitude higher.

With layer thicknesses of 50 cm the pulse shapes become considerably flatter and longer, since the proportion of thermal neutrons of the total flux is significantly higher than for small layer thicknesses. Neutrons already thermalized can remain in the moderator even longer and therefore have a large diffusion time, which is evident by considering the long tail of the pulses. These range up to 4 ms (graphite), 5 ms (D_2O) and 7 ms (beryllium).

After 1 ms the count rate is decreased five orders of magnitude and drops rapidly afterwards. Thus, frame overlap is expected not to occur, even in the case of 50 cm beryllium. If the application of a chopper turns out to be necessary, the neutron yield will presumably not be strongly affected. Of course, these conclusions only hold for not too high repetition rates, e.g., 100 Hz.

Production of Cold Neutrons The next aspect to be investigated is the production of cold neutrons which is conducted with a similar geometry as described above. In addition, a separate layer of liquid deuterium at 19 K is placed behind 30 cm of beryllium which was previously determined as the best possible thermal moderator. As depicted in Fig. 3.7, the spectrum of neutrons leaving the rear-side of the moderator system is shifted to lower energies for increasing thickness. The yield in the energy range of interest from 1.7×10^{-9} MeV to 6.9×10^{-9} MeV corresponding to neutron temperatures between 20 K and 80 K, is at maximum for 12 cm of deuterium. The energy range of interest is indicated by thin gray lines. For layers thicker than 12 cm, the yield decreases due to the absorption reaction $^2H(n, \gamma)^3H$.

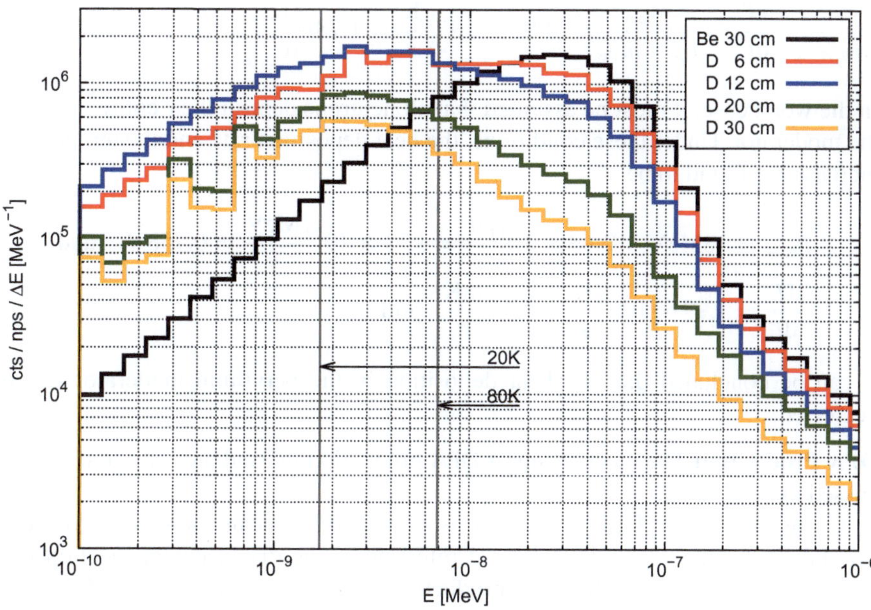

Fig. 3.7 Spectrum of neutrons with an initial energy of 18 MeV behind a layer of 30 cm beryllium and a second layer of liquid deuterium with varying thickness

3.2.2 Heating and DPA Production in the Thermal Moderator

The investigations in this section are carried out for a simple geometry encompassing a spherical beryllium moderator, whereby the converter and thus the source definition for the primary neutrons is located on the surface. First, the heat deposition and the DPA production caused by neutron moderation are calculated. Subsequently, an estimate of the heat deposition of the primary ions is undertaken for the case that the moderator material works simultaneously as the converter.

Heating The heat deposition is determined by an f6-tally averaged over the whole moderator volume. As a result, 2.14×10^{-5} MeV/g are deposited per initial neutron. Considering the total mass of the beryllium sphere, which amounts to $m_{\mathrm{Be}} = 209\,\mathrm{kg}$, a total of 4.5 MeV is deposited per initial neutron in the moderator.

By taking into account the total neutron source strength as shown in Sect. 3.1 and the conservative approach, that the moderator does not dissipate any heat to the environment, the heat load induced by the moderation of neutrons can be calculated as follows

$$\begin{aligned} P_{\mathrm{heat}} &= 2.62 \times 10^{11}\frac{\mathrm{n}}{\mathrm{s}} \times 4.46\frac{\mathrm{MeV}}{\mathrm{n}} = 1.17 \times 10^{12}\frac{\mathrm{MeV}}{\mathrm{s}} = 0.188\,\mathrm{W} \\ c_p(\mathrm{Be}) &= 1825\frac{\mathrm{J}}{\mathrm{kg\,K}} \\ \frac{\mathrm{d}T}{\mathrm{d}t} &= \frac{P_{\mathrm{heat}}}{c_p \times m_{\mathrm{Be}}} = 4.93 \times 10^{-7}\frac{\mathrm{K}}{\mathrm{s}} = 1.78\frac{\mathrm{mK}}{\mathrm{h}} \end{aligned} \tag{3.1}$$

In the worst case, it must be assumed that all neutrons deposit their entire energy in the moderator. Under these conditions, the heat deposition rate and the temperature increase are found to be

$$\begin{aligned} P_{\mathrm{heat}} &= 18\mathrm{MeV} \times 2.63 \times 10^{11}\frac{\mathrm{n}}{\mathrm{s}} = 4.73 \times 10^{12}\frac{\mathrm{MeV}}{\mathrm{s}} = 0.758\,\mathrm{W} \quad \text{and} \\ \frac{\mathrm{d}T}{\mathrm{d}t} &= 1.99 \times 10^{-6}\frac{\mathrm{K}}{\mathrm{s}} = 7.16\frac{\mathrm{mK}}{\mathrm{h}} \quad \text{and} \end{aligned}$$

It can be concluded that the heat deposition in the beryllium moderator can be neglected.

DPA Production The mechanism of DPA production in irradiated structures is discussed in Sect. 2.2.2.

Using MCNP, the DPA production is estimated by means of volume-averaged flux tallies and an `fm`-multiplier card referring to the displacement reaction type `MT=444` which denotes the DPA production cross section in units of MeV b [4]. The values calculated by MCNP are therefore measured in units of $\mathrm{MeV\,b\,cm^{-2}}$. Thus, a normalization constant is defined in the `fm` card which yields the DPA per initial neutron and which also encompasses Eq. 2.3. The normalization constant reads $10^{-24}\frac{\mathrm{cm}^2}{\mathrm{b}} \times 0.8 \times 10^6 \frac{\mathrm{eV}}{\mathrm{MeV}}/62\,\mathrm{eV} = 1.29032 \times 10^{-20}\,\mathrm{eV}^{-1}$. Accordingly, the `fm` card is written as

```
FM104 1.29032E-20 1 444
```

where 1 denotes the material number of beryllium in the MCNP input.

The DPA production per full power year[1] is calculated as the product of the MCNP tally result and the neutron source strength at Trident to

$$8.96 \times 10^{-26}\frac{\mathrm{DPA}}{\mathrm{n}} \times 2.63 \times 10^{11}\frac{\mathrm{n}}{\mathrm{s}} = 2.36 \times 10^{-14}\frac{\mathrm{DPA}}{\mathrm{s}} = 7.43 \times 10^{-7}\frac{\mathrm{DPA}}{\mathrm{fpy}}.$$

It is evident that the DPA production by neutrons is irrelevant with regard to the integrity of the beryllium moderator.

Heating by Deuterons In the two previous sections it turned out, that heat deposition and DPA production by neutrons in a beryllium-based moderator can be neglected. This is not the case for the converter which is impinged by the high-energetic deuterons. In this section the heat deposition in a beryllium bulk converter is investigated based on the spectrum of the measured deuterons, see Fig. 3.2. Contrary to Trident, however, the deuterons are completely forward directed so that the simulation can be considered conservative.[2] The total energy deposition is calculated by MCNP to 29.79 MeV per initial deuteron which corresponds to 1.5×10^{14} MeV/s = 24 W assuming an operation at 10 Hz. Considering Eq. 3.1 the mean temperature increase of the spherical beryllium converter-moderator system amounts to

$$\frac{\mathrm{d}T}{\mathrm{d}t} = \frac{P_{\mathrm{heat}}}{c_p \times m_{\mathrm{Be}}} = 6.29 \times 10^{-5}\frac{\mathrm{K}}{\mathrm{s}} = 0.23\frac{\mathrm{K}}{\mathrm{h}}.$$

It can be concluded that heat deposition by neutrons and the primary deuterons as well as DPA production are easily manageable in the regime which prevails at Trident operating at 10 Hz.

3.2.3 Source Definition Based on Trident Results

The preliminary investigations were carried out based on simplified source definitions. In order to be able to precisely design and optimize the moderator system of a laser-driven neutron source, all measurement results for the neutron production at Trident with a 480 nm thick CD foil are used.

For this aim, the values of the measured neutron spectrum (Fig. 3.3a), the pulse shape (Fig. 3.3b) and the angular distribution (Fig. 3.4) are digitized with aid of

[1]One fpy corresponds to the total time during which a device or a facility is operated at full power for an entire year.

[2]In reality, a scenario like this could be created by enclosing the target foil by converter material as much as possible.

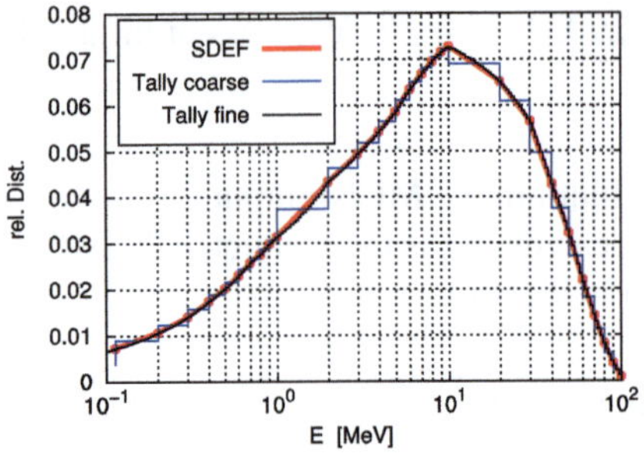

(a) Spectrum derived from the experimental results in figure 3.3a.

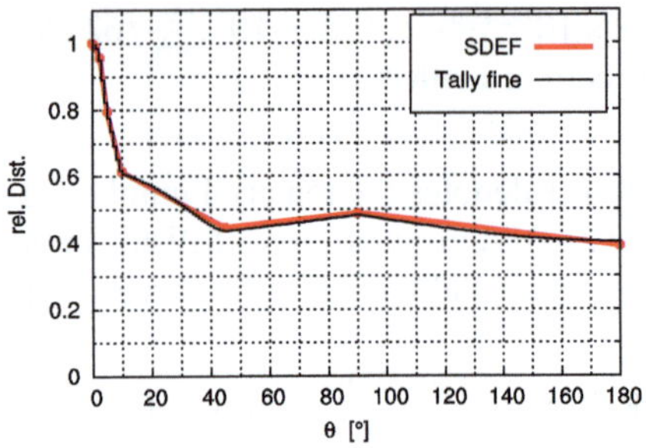

(b) Angular distribution with respect to the deuteron beam axis derived from the experimental results in figure 3.4.

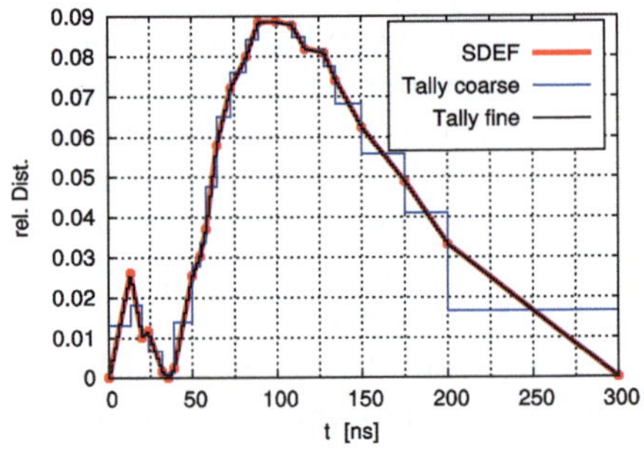

(c) Pulse structure derived from the experimental results in figure 3.3b.

Fig. 3.8 MCNP source distributions of the primary neutrons based on the measurements at Trident with deuterated PE film in the BOA regime. The implementation of the source definition is validated by means of MCNP simulations with corresponding tallies

the software GSYS2.4 [5] and are implemented as a source definition for MCNP. A visualization of the datapoints used in the source definition is shown in Fig. 3.8. According to [6], the data points are implemented as vertices of a piecewise linear function and not as a histogram, which corresponds to the red curve. A simple simulation geometry is applied for validation, a spherical surface is positioned around the source, on which the spectrum and the temporal pulse structure are determined by surface flux tally. The angular distribution, on the other hand, is determined with a cylindrical mesh-based tally. Due to the rotational symmetry of the source definition, it is sufficient to apply a thin disk of the cylindrical mesh tally which contains the source point. The results of the tallies for validation purposes are represented by the black and the blue curve. The blue curve corresponds to a tally with a coarse energy and time binning and the black curve to a fine binning. There is a very good agreement between the `SDEF` input data and the tally results. The implementation of the source definition is therefore considered reliable and correct. In all subsequent simulations, this source definition is used unless otherwise specified.

3.3 Setup and Design of the Moderator

In this section the results of the simulation studies for the design of a moderator system for the generation of cold neutrons with high brilliance are elucidated. The assembly consists of a thermal moderator, in which the fast primary neutrons are thermalized, and a small-volume cold moderator. According to the work of [7, 8] the brilliance of cold neutrons can be significantly increased by using small-volume, rod-shaped or disk-shaped cold moderators.

For the simulations, the source definition of the primary neutrons generated at Trident derived in the previous section is placed directly on the surface of the moderator. According to the results of the preliminary studies, beryllium is chosen as material for the thermal moderator. First, the external shape of the thermal moderator is optimized with respect to the maximum value of the thermal flux. Subsequently, the material and dimensions of cold moderators, which are embedded in the thermal beryllium moderator, are investigated. In the following simulations, neutrons with an energy $E \leq 30\,\text{MeV}$ corresponding to a temperature of 348 K are considered thermal.

3.3.1 Moderator Shape and Geometry

To identify the optimum shape of the thermal moderator, the value and the location of the maximum thermal flux is determined in dependence of the volume of the moderator by means of a mesh-based tally which has a resolution of 1 cm in each direction. The simulations are performed for the following geometries: sphere, cylinder, cone, ellipsoid, and cap (cylinder with a superimposed half-sphere) . Since the volume of all configurations except the sphere depends on two parameters, radius r and height h, both parameters must be varied independently. The results are shown in Fig. 3.9. For each point in the diagram only the flux value of the optimum combination of r and h is shown.

The maximum value of the thermal flux increases rapidly up to a volume of approximately $100\,\text{dm}^3$. From 200-300 dm^3, no remarkable increase in the maximum flux value can be observed. For very large volumes the maximum thermal flux becomes saturated where as many neutrons are absorbed or escape the moderator, as are backscattered to the center of the moderator.

The highest flux is achieved in the cap configuration, which is closely followed by cylindrical and conical designs. These configurations have nearly identical saturation values of $1.73 \times 10^8\,\text{cm}^2\,\text{s}^{-1}$. In ellipsoidal moderators maximum flux values of $1.6 \times 10^8\,\text{cm}^2\text{s}^1$ are observed. The simplest moderator geometry, the sphere, also provides the lowest thermal flux with $1.37 \times 10^8\,\text{cm}^2\text{s}^-1$. Interestingly, the saturation values for all geometries are not identical, although the same material is used. This can be explained by the fact that the source of the primary neutrons, i.e., the converter, is located on the surface of the moderators. Due to the angular dependence, see Figs.

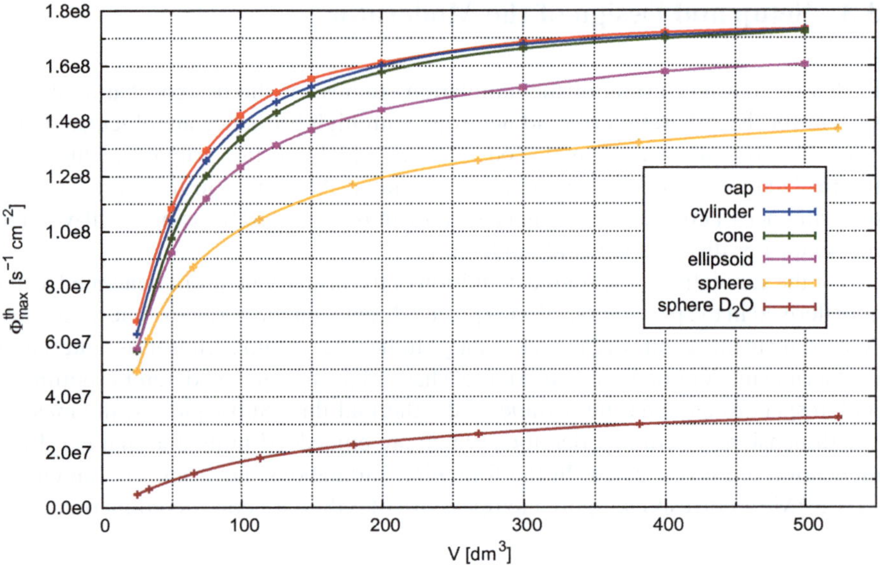

Fig. 3.9 Maximum value and relative error of the thermal flux for different shapes of a beryllium moderator depending on their volume. For spherical moderators also heavy water is investigated. Values are normalized to the neutron yield at Trident assuming an operation at 10 Hz, which corresponds to 2.62×10^{11} n/s

3.4 and 3.8b, curved surfaces, i.e., the sphere and the ellipsoid, can not capture as many primary neutrons as planar surfaces which are the bases of cap, cylinder and cone.

To refer again to the material selection, which is based on the preliminary studies shown in Sect. 3.2.1, a spherical geometry with heavy water is investigated. Despite the very good moderating ratio of heavy water, a thermal flux which is not nearly as high as in beryllium can be achieved. For large volumes values of only $3.24 \times 10^7\ \mathrm{cm^{-2}s^{-1}}$ are observed. It can be seen that the (n, 2n) cross section of beryllium has a large impact on the thermal flux, which is not taken into account in the theoretical calculation of the moderating ratio, cf. Sect. 2.3.2. The energy of the primary neutrons generated at Trident are comparatively high, as indicated by the spectrum in Fig. 3.3a. According to the shape of the (n, 2n) cross section, these high-energy neutrons can be effectively multiplied by (n, 2n) and (n, 3n) reactions and simultaneously decelerated in beryllium. For the further simulations, a cylindrical beryllium moderator with a volume of $V = 125\,\mathrm{dm}^3$ is used with optimum dimensions of $r = 31$ cm and $h = 41$ cm. Above this volume, the gain in thermal neutrons at the center of the moderator is too small to justify the additional material costs. The position of the maximum thermal flux is given by (0, 0, 15.7) cm.

3.3.2 Converter Position

The source definition of the deuterons is based on the results of the Trident experiments. For the sake of simplicity, the emission direction of the deuterons is assumed to be limited to a cone with half an apex angle of $\theta_{\mathrm{apex}} = 15°$. According to Fig. 3.2a, the deuteron flux is approximately constant in this angular range. The spectrum of the deuterons is extracted from diagram 3.2b using the software GSYS2.4 [5] and implemented in the source definition.

The simulation model is depicted in Fig. 3.10 and consists of a cylindrical beryllium moderator which is equipped with a conical inlet accounting for the angular distribution of the deuteron source. The deuteron source is positioned in 1 cm distance to the moderator on the axis of symmetry. Accordingly, the openings radius is set to $r_{\mathrm{c}} = 0.3$ cm. In this configuration, the first layer of the moderator behind the base of the conical inlet serves as a converter in which neutrons are produced. In a parameter study, the height h_{c} of the conical inlet is varied between 1 cm and 16 cm, whereby, the inner radius R_{c} is also varied for each value of h_{c}.

In Fig. 3.11, the maximum thermal flux in the moderator is depicted in dependence on the height h_{c} for corresponding optimum value of R_{c}. As expected, the optimum flux value is always achieved for the combination of h_{c} and R_{c}, where the cone has a half opening angle of 15°. It can be clearly seen that the maximum thermal flux in the moderator can be considerably increased as the converter is moved deeper into the moderator. As the length of the inlet increases, the thermal flux increases almost linearly, until the curve gradually decreases above 8 cm. The maximum is reached at 14 cm with a flux value of $1.13 \times 10^7 \mathrm{cm}^{-2}\mathrm{s}^{-1}$. For larger h_{c}, the maximum achievable flux decreases again being caused by the fact that the moderator thickness

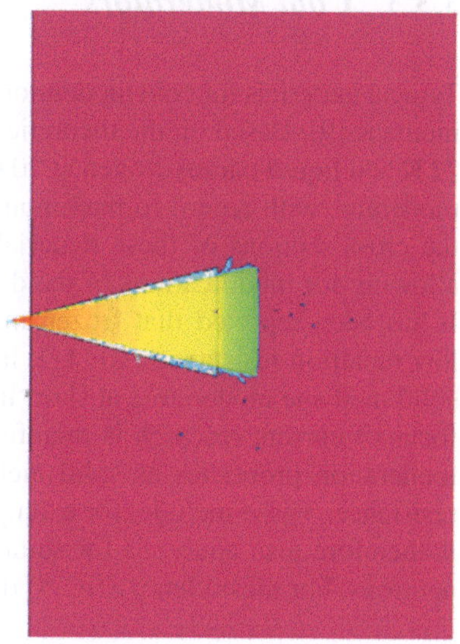

Fig. 3.10 MCNP model of a cylindrical beryllium moderator with a conical inlet ($h_{\mathrm{c}} = 10$ cm and $\theta_{\mathrm{apex}}/2 = 15°$) for the primary deuterons produced at Trident. Accordingly, the first layers of beryllium at the base of the inlet cone serves as the neutron-producing converter. The length of the inlet as well as the radius of the cone's base are varied in the parameter study. The flux distribution of the primary deuterons can also be seen

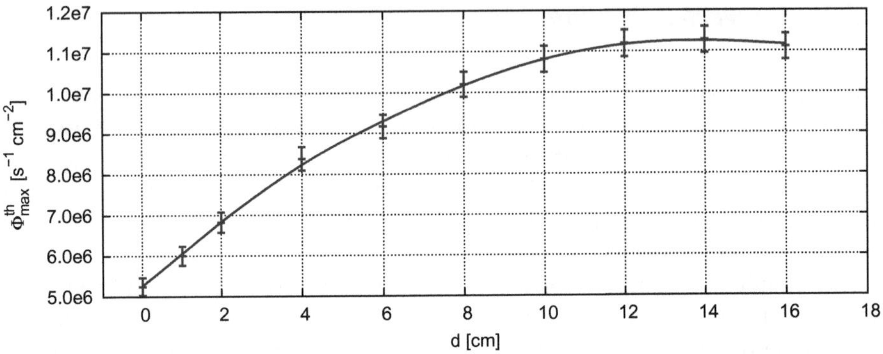

Fig. 3.11 Maximum thermal flux and relative error inside the moderator in dependence to the length of the conical inlet for the primary deuterons. As the radius is also varied for each length, only the optimum flux values are shown

amounts to 31 cm and above $h_c = 15.5$ cm more moderator material is behind the converter area than in front of it.

The results show that for larger h_c, the rear layers of the moderator serve as a reflector for backwardly directed neutrons. As shown in the previous section, the increase in thermal flux is limited by increasing the moderator or by using a reflector. It is therefore advisable to install the converter as far as possible into the moderator even in larger moderators in order to minimize the leakage of primary neutrons in the reverse direction.

3.3.3 *Cold Moderators*

Typical materials for cold moderators are liquid hydrogen, liquid deuterium and solid methane [9]. Based on the theoretical considerations in Sect. 2.3.2 solid methane at 22 K and liquid parahydrogen at 20 K are taken into account for optimizing the cold moderator with regard to maximum yield and brilliance. According to Sect. 2.5.4 the cross sections of these materials are called in MCNP by means of `MT`-cards allowing for modifications of the default cross sections by additional $S(\alpha, \beta)$ data. It has been reported that frozen methane offers a higher effectiveness, but has a low radiation resistance [10, 11]. It is therefore necessary to replace or regenerate solid methane moderators at short intervals because of the high neutron fluxes. The focus of current research is therefore on aromatic compounds which have similar moderation properties as solid methane but offer a significantly higher radiation resistance. These include, for example, mesitylene and xylene [12, 13]. The former is therefore also analyzed for suitability and efficiency for the production of cold neutrons. For mesitylene, $S(\alpha, \beta)$ data is available for a temperature of 32 K [14].

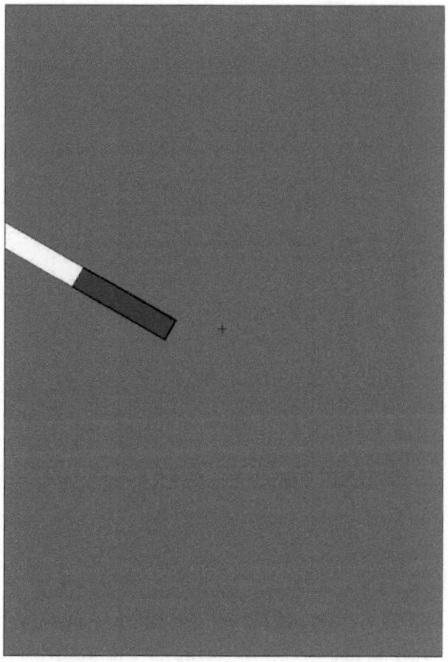

Fig. 3.12 Simulation model for determining the optimum dimensions of the cold moderator (dark gray) embedded in the thermal moderator (light gray) for different materials concerning the spectrum of primary neutrons produced at Trident. Here, the configuration with parahydrogen, $r = 1$ cm and $l = 10$ cm is shown

The cylindrical cold moderator is placed in the thermal beryllium moderator in such a way that its lower edge is located at the thermal flux maximum. In order to minimize the background of fast neutrons, the position or the angle of incidence of 150° is taken as an example. The geometry for parahydrogen with $r = 1$ cm and $l = 10$ cm is presented in Fig. 3.12. Optimum values for length and radius of the cold moderator are determined in terms of the highest yield of cold neutrons with $E < 7\text{meV} \mathrel{\widehat{=}} 81\text{K}$ at the outwardly directed surface of the cold moderator. The results are depicted in Fig. 3.13 and summarized in Table B.1 in the appendix. As a supplement, the dependence of the cold flux on the length of the moderators are depicted in Fig. B.1 in the appendix for the optimum radius of each material.

All materials have in common that the flux of cold neutrons increases with the radius of the moderator and then decreases after exceeding an optimum value of 2–3 cm. For each data point in the table and the diagram, the optimum length is considered for each radius as a result of a two-dimensional parameter study. Due to the shape of the cross section of parahydrogen, which implies that the mean free path for cold neutrons is much larger than for thermal neutrons $L_c \gg L_t$, optimum moderator shapes of parahydrogen tend to be rod-shaped with $l = 5$ cm and $r = 1.5$ cm, whereas solid methanes best configuration is more compact with $l = 2.5$ cm and $r = 1.5$ cm. As mentioned before, methane provides the best performance for cold neutrons. The best possible cold neutron flux is about 36% higher than the best value of parahydrogen and even 133% above that of mesitylene.

Spectra for different radii are shown in Fig. 3.14. Apart from the fact that solid methane delivers a higher cold neutron yield, the maximum yield in parahydrogen is achieved at 8 MeV corresponding to 90 K, while solid methane shows a smooth maximum of the cold neutron flux between 20 and 80 K. This can be understood

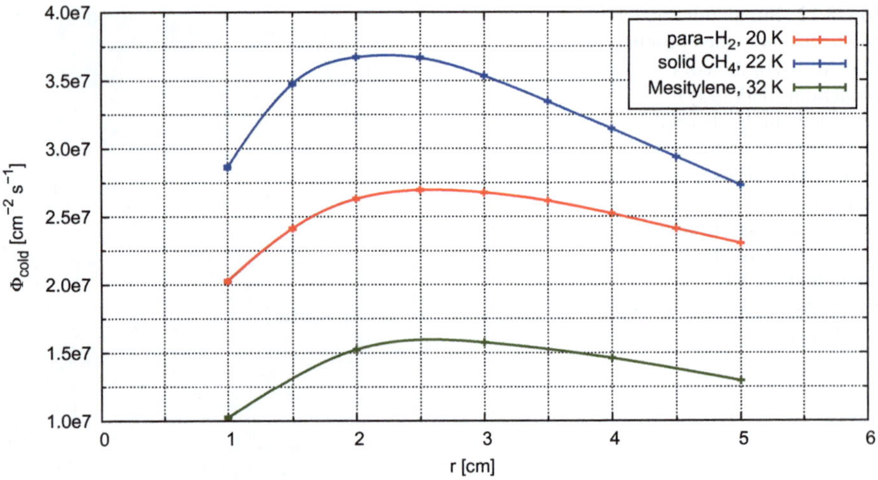

Fig. 3.13 Flux of cold neutrons with $E < 7\,\text{meV} \mathrel{\widehat{=}} 81\,\text{K}$ on the outwardly directed cold moderator surface depending on the radius. For each radius value the optimum length is taken into account. Values are given in terms of the neutron production at Trident operating at 10 Hz

by considering the steep slope around 100 K of the parahydrogens scattering cross section given in Fig. 2.8b. The maximum flux for mesitylene appears at 90 K, as in the case of parahydrogen. With the same radius, however, the spectrum in the cold range is only half as high. Even taking the fact into account that the present cross section data of mesitylene is evaluated for a temperature which is 10 K higher than that of parahydrogen, it can not be expected that with colder mesitylene higher cold fluxes can be obtained than with parahydrogen or methane.

In order to be able to analyze the directional characteristics of the emitted cold neutrons, the brilliance for the materials under consideration is determined. Results for $r = 1, 2$ and 5 cm are depicted in Fig. 3.15. As explained above, the brilliance of cold moderators based on parahydrogen can be increased by reducing the extension of the moderator to a value close to the thermal neutron scattering length $L_t \approx 1$ cm in at least one dimension. This can be confirmed by the simulations. In the forward direction with $\theta \leq 5°$, a parahydrogen moderator with $r = 1$ cm results in a higher cold flux than with larger radii. The shape of the brilliance is, however, also much steeper, and thus for angles above 5° the cold neutron flux is higher for $r = 2$ cm. It should be noted that current supermirrors can only handle cold neutrons up to a critical angle of about $\theta_{\text{crit}} \approx 4°$, see Sect. 2.4. The shape of the brilliance for solid methane and mesitylene are similar. However, the absolute values correspond to the angle-integrated cold flux values from Table B.1 and Fig. 3.13. For a methane moderator with $r = 2$ cm a similar brilliance is achieved as with the smallest parahydrogen moderator. For all other radii, the brilliance is significantly lower. In the best case the peak brilliance of mesitylene is 60% below the peak brilliance of parahydrogen and methane.

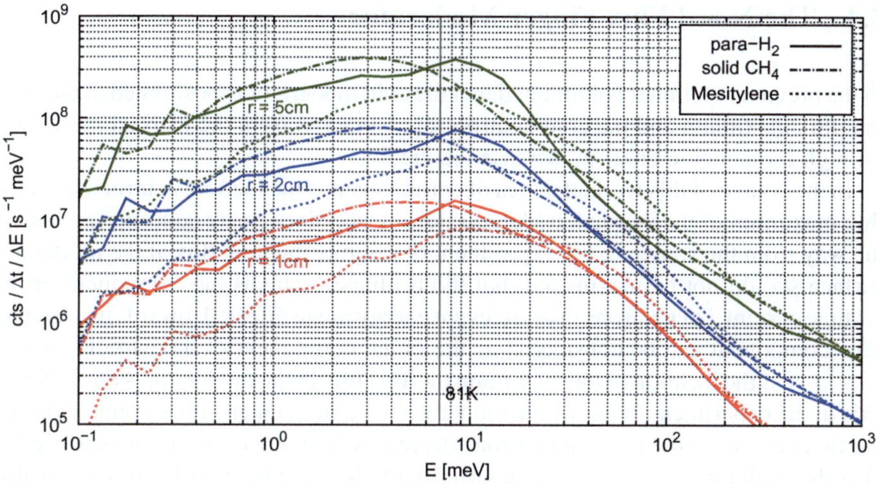

Fig. 3.14 Spectrum of neutrons on the outwardly directed cold moderator surface for different radii and materials. For each curve the optimum length of the moderator is taken into account, see Table B.1. Values are given in terms of the neutron production at Trident operating at 10 Hz. For the sake of clarity, and to emphasize that indeed the flux of cold neutrons is maximal at 2–3 cm, but the total yield of cold neutrons increases with the radius, the spectra are not normalized to the area

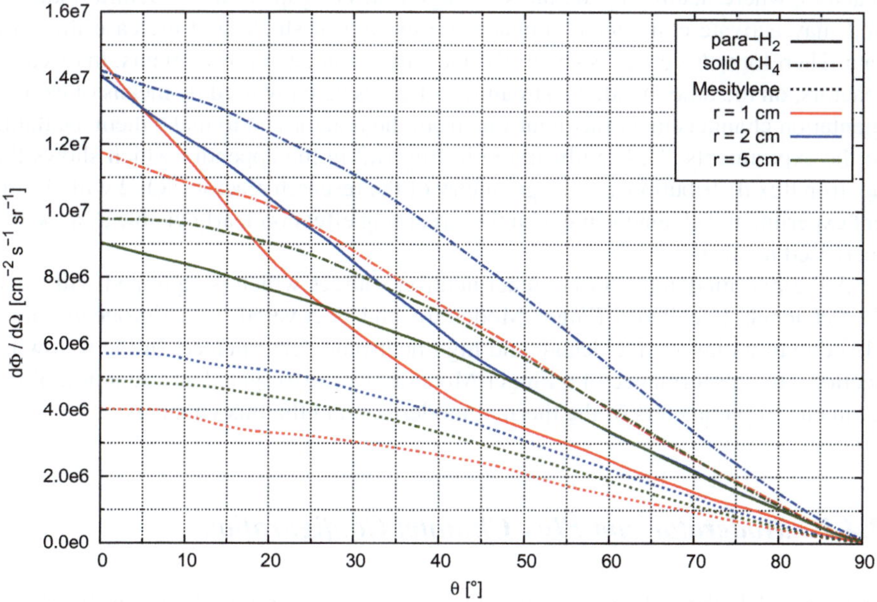

Fig. 3.15 Brilliance of cold neutrons with $E < 7\text{meV} \,\widehat{=}\, 81\,\text{K}$ on the outwardly directed cold moderator surface for different radii and materials. For each curve the optimum length of the moderator is taken into account. Values are given in terms of the neutron production at Trident operating at 10 Hz

3.4 The Novel Flux Finger Moderator

In the previous section it is explained that the cold moderator is positioned at the thermal flux maximum and the cold neutrons are extracted from the thermal moderator through a channel in order to increase the yield.

In the course of this work, the idea of extracting also the thermal neutrons from the flux maximum of the thermal moderator developed. Since many neutrons are lost on the way from the source to the experiment or sample, the total flux at the moderator surface should not be maximized, but the brilliance. This is the component of the particle current $\vec{j}$ of neutrons in the relevant energy range directed towards the sample. This component can be considered as the radial component of the particle current $j_r \equiv \vec{j} \cdot \vec{r}$ if the location of the maximum thermal flux in the moderator is on the beam axis of the experiment. If the whole moderator volume is taken into account, not only the flux density of the thermal neutrons decreases from the center to the surface but also the radial component of the thermal particle current. In order to extract this component from the flux maximum, evacuated flux channels are implemented into the moderator. Neutrons originating from the thermal flux maximum with the "right" direction can enter the instrumentation without further collisions and thus without changing their direction and can be transported to the sample.

This effect commonly designated as the *streaming effect* is known from nuclear reactors, where neutrons penetrates through tubes or gaps in the structural material and may activate components outside the biological shield and increase the dose rate. Therefore, this effect is generally undesired in nuclear power plants. In research reactors, on the other hand, the streaming effect is actively used by means of experimentation channels to extract neutrons from the reactor and to make them available to the experiments. This is illustrated in Fig. B.2 in the appendix which shows the neutron flux distribution and the geometry of the research reactor FRJ–2 with different experimental channels in which the streaming effect is clearly visible as a locally increased neutron flux.

A thermal moderator for compact neutron sources which is equipped with such flux channels is called a Finger Moderator and is covered by a European patent [15] due to its innovative design and high neutronic performance. In the following sections, the performance of a Finger Moderator is investigated as a function of the geometrical parameters of the thermal moderator and the flux channels.

3.4.1 *Moderator and Flux Channel Configurations*

In Sect. 3.3.1, the influence of the outer shape of moderators based on beryllium and heavy water on the maximum thermal flux is examined. The more important parameter is, however, the flux or the brilliance of thermal neutrons, which leave the moderator and are actually available to the experiment. Therefore, the thermal flux at the channel exit is determined for different configurations of the Finger Moderator.

For this purpose spherical and cylindrical shapes as well as the cap configuration and different materials are taken into consideration. Additionally, spherical moderators which are equipped with a reflector are investigated. Beryllium and heavy water are taken into account as moderating materials whereas beryllium and graphite are used as reflector materials. The simulation models with a flux channel with radius $r = 1\,\text{cm}$ are shown in Fig. 3.16. For each moderator configuration, a parameter study is performed in which the angle of incidence φ of the flux channel is varied. This angle is defined with respect to the axis of the deuteron beam.

Results of the parameter study are depicted in Fig. 3.17. When comparing beryllium-based moderators with different shapes, it is noticeable that for spherical moderators (blue curve) the thermal flux at the channel exit is lower for most angles than for the cylinder (red curve) and the cap (green curve). The thermal flux of the spherical moderator is higher only at angles of incidence where the flux channel would sweep the edges of the cylindrical moderator. This is the case for $\varphi = 50°$ and $\varphi = 115°$. At these positions, the flux channel is significantly longer so that less neutrons from the center of the moderator reach the channel exit. The shape of the curve of the thermal flux in the cap moderator is identical to that of the cylinder above an angle of $\varphi = 100°$. For smaller angles, the shape of the flux is flatter, as the channel does not sweep a further edge.

The orange curve denotes the thermal flux on the surface of a bulk beryllium-based moderator without flux channels. The values are determined at an area that corresponds to the channel exit of the other configurations and are a factor 1.5–2 lower compared to the cylindrical moderator with flux channels.

The magenta curve shows the thermal flux for a spherical D_2O moderator. The values are lower by a factor of 2.4 ($\varphi = 0°$) to 6.2 ($\varphi = 180°$) than for the spherical beryllium moderator. The results shown in Sect. 3.3.1, Fig. 3.9 on the maximum thermal flux in the moderator center can therefore be confirmed.

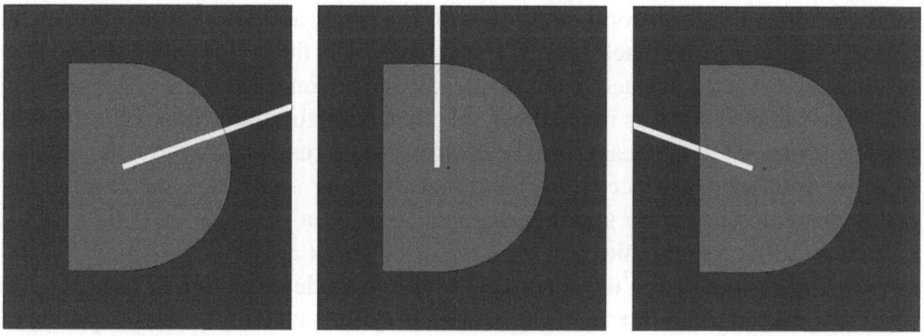

(a) 20° **(b)** 90° **(c)** 160°

Fig. 3.16 Simulation model for investigating the thermal flux at the channel exit in dependence to its angle of incidence with respect to the axis of symmetry of the moderator. Here, the configuration of a cap-shaped beryllium moderator with graphite reflector (thickness 20 cm measured from the flat surfaces) for different angles of incidence of the flux channel

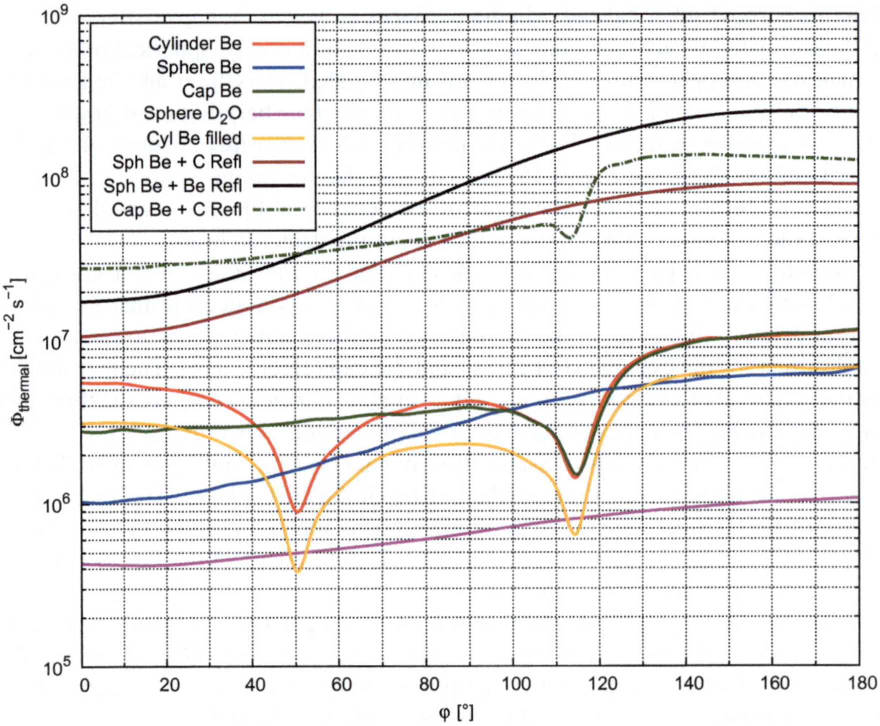

Fig. 3.17 Flux of thermal neutrons measured at the flux channel exit in dependence of the channel inclination φ for different moderator configurations at Trident operating with 10 Hz. The values of the two configurations with a reflector were determined at the moderator surface and not at the reflector surface

The results of the two configurations with a graphite and a beryllium reflector are given by the brown and black curves, respectively. The flux values are determined at the channel exit at the surface of the moderator and not on the surface of the reflector, i.e., at the same position as in the other spherical configurations. In a real setup, this would correspond to the beam lines or neutron guides reaching through the reflector up to the moderator surface. It is found that by using a reflector, the thermal flux at the channel exit can be significantly increased by a factor of 10.3 ($\varphi = 0°$) to 13.4 ($\varphi = 180°$) for graphite and by a factor of 16.9 ($\varphi = 0°$) to 37.3 ($\varphi = 180°$) for beryllium compared to the spherical beryllium moderator without reflector. The behavior of the thermal flux for a cap-shaped beryllium moderator with a cylindrical graphite reflector (green dashed curve) is also shown. Accordingly, the flux is at least one order of magnitude higher for all angles of incidence than for the cap-shaped beryllium moderator without a reflector. It can also be seen that the dips of the flux at the edges is markedly weaker.

For each of the configurations, the tendency can be observed that the thermal flux at the channel exit is higher for large φ (opposite direction to the incoming deuteron beam) than for low φ. This, and the fact that the edges of the cylinder and the cap lead to a significant drop in thermal flux, show that this is essentially a longitudinal effect of the channel.

3.4.2 Shape of the Flux Channels

In addition to the position of the flux channels, their shape is also investigated in view of higher neutronic performance. For a cylindrical beryllium moderator with a volume of $125\,\mathrm{dm}^3$, the radii of the flux channels with $\varphi = 0°$, $10°$, $150°$ and $160°$, corresponding to the red curve in Fig. 3.17, are varied while determining the thermal flux at the channel exit. Results are depicted in Fig. 3.18 and show that for increasing radius, the thermal flux increases rapidly to a certain maximum value and then decreases. The maximum value amounts to $\Phi = 1.27 \times 10^7 \mathrm{cm}^{-2}\,\mathrm{s}^{-1}$ and $\Phi = 1.15 \times 10^7 \mathrm{cm}^{-2}\,\mathrm{s}^{-1}$ for $\varphi = 160°$ and $\varphi = 150°$ and is reached at $r = 3.5\,\mathrm{cm}$ and $r = 5\,\mathrm{cm}$, respectively. For $\varphi = 10°$ and $\varphi = 0°$, the maximum flux can be achieved with $r = 4.8\,\mathrm{cm}$ with values of $\Phi = 1.27\,\mathrm{cm}^2\,\mathrm{s}^{-1}$ and $\Phi = 1.15\,\mathrm{cm}^{-2}\,\mathrm{s}^{-1}$. The values for $r = 1\,\mathrm{cm}$ are in accordance with the results in Fig. 3.17 confirming the consistency of the different simulations.

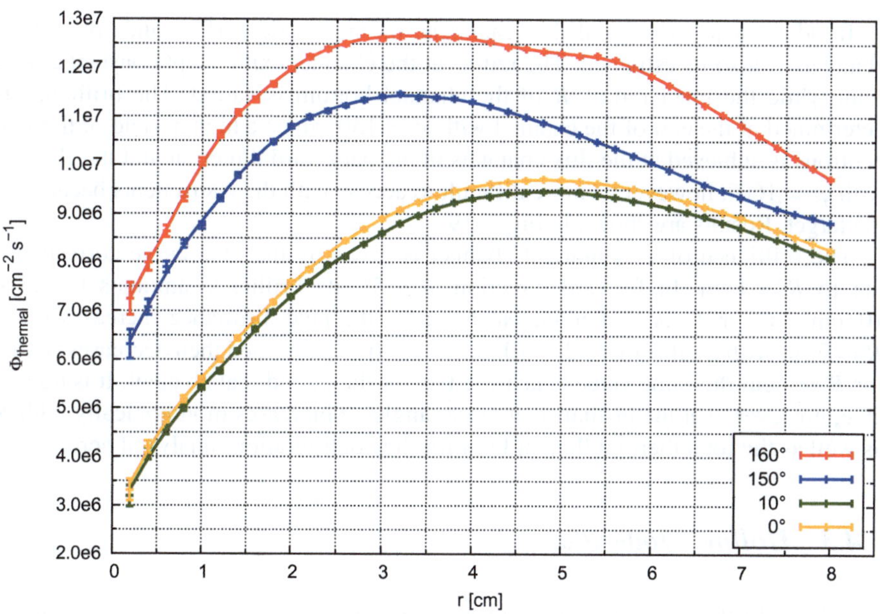

Fig. 3.18 Flux of thermal neutrons measured at the flux channel exit in dependence of the channel radius for different angles of incidence φ of the channel. Values are normalized to Trident operating at 10 Hz

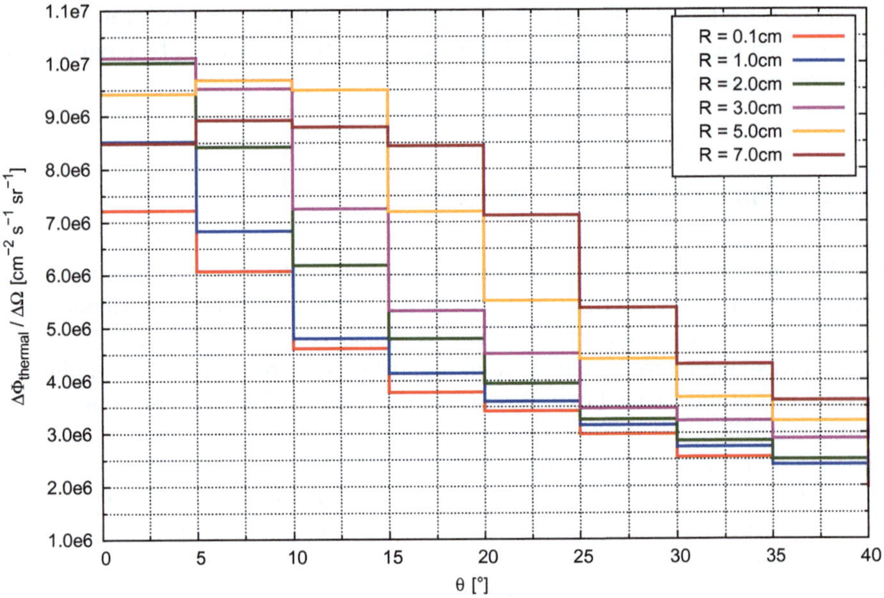

Fig. 3.19 Brilliance of thermal neutrons at the exit of a conical flux channel with $\varphi = 150°$. The radius at the channel exit is kept constant with $r = 1\,\text{cm}$ while the inner radius R is varied

In addition to cylindrical flux channels, the neutronic performance of conical channels is also investigated. The radius at the channel exit is kept constant at $r = 1\,\text{cm}$ while the inner radius R of the channel bottom is varied. The brilliance is determined at the exit of the channel with $\varphi = 150°$ since a higher yield of thermal neutrons is achieved at the flux channels in the backward direction and the channel with $\varphi = 160°$, however, is more difficult to access with the instruments because of the target. Results are shown in Fig. 3.19.

The brilliance of a cylindrical channel is given by the blue curve corresponding to $R = 1\,\text{cm}$. In forward direction, $\theta \leq 5°$, its brilliance amounts to $8.5 \times 10^6\,\text{cm}^{-2}\,\text{s}^{-1}\,\text{sr}^{-1}\,\mu\text{A}^{-1}$. For increasing radii, the peak brilliance increases up to $R = 3\,\text{cm}$ with a value of $1.02 \times 10^7\,\text{cm}^{-2}\,\text{s}^{-1}\,\text{sr}^{-1}\,\mu\text{A}^{-1}$ and then decreases so that for $R = 7\,\text{cm}$ the same value is achieved as for the cylindrical channel. It is noticeable, however, that the brilliance at larger angles θ increases monotonically with R as well as the total thermal flux, at least for the configurations analyzed here.

3.4.3 *Hollow Moderator*

In order to further investigate the streaming effect additional simulations are carried out in which a cavity is implemented into the moderator at the location of the maximum thermal flux which occurs at a depth of 17.7 cm in beam direction. For the sake of simplicity the moderator and the cavity are of spherical shape with a radius of

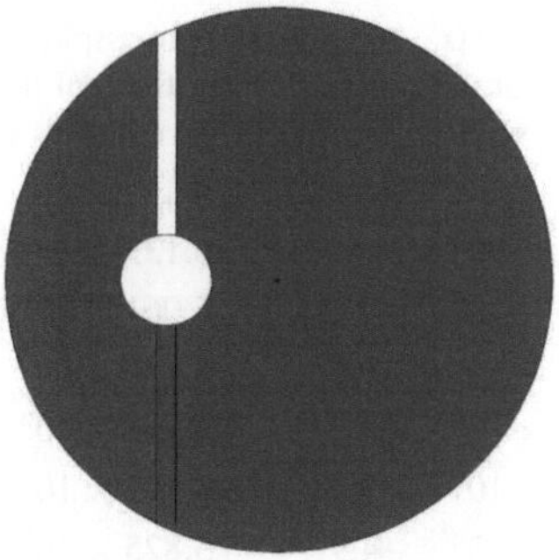

(a) Channel inclination 90° and cavity radius $r_{cav} = 3$ cm.

(b) Channel inclination 150° and cavity radius $r_{cav} = 1.5$ cm.

Fig. 3.20 Simulation models for determining the dependence of the thermal flux at the channel exit in dependence of the radius of the spherical cavity located at the thermal flux maximum inside a spherical beryllium moderator with $r = 30$ cm

$r = 30$ cm which corresponds to a volume of $V = 113.1\,\text{dm}^3$. A flux channel with radius $r = 1$ cm is inserted and the thermal flux is calculated at its exit in dependence to the radius r_{cav} of the cavity in a range between 0 cm and 5 cm. A value of 0 corresponds to the case of the flux channel extending to the point of maximum thermal flux and absence of the cavity. Additionally, different angles of inclination of the flux channel are investigated with values of $\varphi = 0°, 90°$ and $150°$. The geometry for $r_{cav} = 5$ cm and $\varphi = 90°$ is depicted in Fig. 3.20b and for $r_{cav} = 2$ cm and $\varphi = 150°$ in Fig. 3.20a, respectively. The results of the simulation study are presented in Table 3.1. For each flux value, the percental deviation is shown in comparison to the case without cavity. It is clear that the thermal flux at the channel exit can not be significantly influenced with small cavities. There is no tendency to observe, as the values are sometimes higher and sometimes lower for different radii r_{cav} and angles φ which is due to the statistical uncertainty of the simulation results.

3.4.4 Multiple Flux Channels

The question arises whether several flux channels in one moderator system mutually interfere with one another. According to Fig. 3.21, two models are considered in which four flux channels are implemented. In the first model, the flux channels have a cylindrical shape with $r = 1$ cm and in the second model a conical shape, with an inner radius of $R = 4$ cm. Two of the flux channels are arranged at an angle of incidence of $\varphi = 10°$ and $-10°$, respectively, and the other two at an angle of incidence of $\varphi = 150°$ and $-150°$, respectively, defined in direction of the incident deuterons.

Both the flux and the brilliance of thermal neutrons are determined at the channel exits and compared to models in which only one flux channel is present. The flux values are summarized in Table 3.2. In addition, the loss of thermal neutrons is

Table 3.1 Thermal flux the exit of channels with $\varphi = 0°$, $90°$ and $150°$ in a spherical beryllium moderator with a spherical cavity of varying radius at the position of the thermal flux maximum. The values are compared to the values of the model without cavities in the columns entitled *gain*

r_{cav} (cm)	$\Phi_{th}^{0°}$ ($cm^{-2}\,s^{-1}$)	Gain (%)	$\Phi_{th}^{90°}$	Gain (%)	$\Phi_{th}^{150°}$	Gain (%)
0.0	4.824×10^6	–	5.163×10^6	–	6.133×10^6	–
1.5	4.993×10^6	3.5	5.158×10^6	−0.1	6.061×10^6	−1.2
2.0	4.884×10^6	1.2	5.028×10^6	−2.6	6.150×10^6	0.3
3.0	4.716×10^6	−2.2	5.207×10^6	0.9	6.148×10^6	0.2
4.0	4.749×10^6	−1.6	4.983×10^6	−3.5	6.113×10^6	−0.3
5.0	4.891×10^6	1.4	4.909×10^6	−4.9	5.825×10^6	−5.0

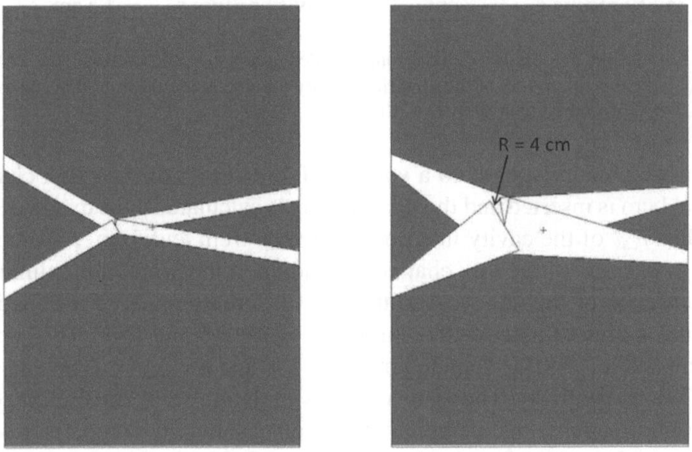

Fig. 3.21 Simulation models for the investigation of the mutual influence of several flux channels. Both cylindrical channels with $r = 1$ cm as well as conical channels with $r = 1$ cm and $R = 4$ cm are considered

determined for the case of four channels compared to a single one. For symmetry reasons only the two channels with 10° and 150° are listed.

For four cylindrical channels, the loss of thermal neutrons at the channel exit is negligible with 1.2% and 1.5%, respectively; especially if the relative error of the flux values of approximately 0.8% is taken into account. In the case of conical channels, however, the deviation is significantly higher. For 10°, the flux decreases by 6.6% compared to a single channel, and for 150° by 15.5%. Considering the geometry in Fig. 3.21, it becomes clear that with conical channels, which are all positioned at the flux maximum, a considerable amount of moderator material is left out, which is responsible for the effect shown above. This findings are substantiated by the results from Sect. 3.4.3, in which is examined how a cavity in the moderator influences the performance.

Apart from the total thermal flux at the channel exit, the thermal flux in the forward direction should also be considered, since only these neutrons are available for the distant experiment. The brilliance at the channel exits is shown in Fig. 3.22 in which

Table 3.2 Thermal flux Φ_{th}, peak brilliance $\Delta\Phi_{th}^{fw}/\Delta\Omega$ in forward direction ($\theta = 1°$) and their relative errors at cylindrical and conical channel exits in a model with multiple channels and models with single channels. The column *Loss* shows the loss of thermal neutrons when using multiple flux channels compared to single channels

Flux channel	Φ_{th} (cm^{-2} s^{-1})	σ_{rel} (%)	Loss (%)	$\Delta\Phi_{th}^{fw}/\Delta\Omega$ (cm^{-2} s^{-1} sr^{-1})	σ_{rel} (%)
Cylinder multi 10°	5.65×10^6	1.03	−1.5	1.02×10^7	4.6
Cylinder single 10°	5.73×10^6	1.02		1.12×10^7	4.4
Cylinder multi 150°	1.05×10^7	0.78	−1.2	9.82×10^6	4.5
Cylinder single 150°	1.06×10^7	0.78		1.09×10^7	4.7
Cone multi 10°	7.09×10^6	0.93	−6.6	6.43×10^6	5.6
Cone single 10°	7.59×10^6	0.89		8.32×10^6	5.1
Cone multi 150°	1.12×10^7	0.75	−15.5	7.07×10^6	5.6
Cone single 150°	1.33×10^7	0.69		1.12×10^7	4.5

the solid lines indicate the results of the simulations with multiple flux channels and dashed lines with a single channel. It can be seen that the brilliance for multiple channels and single channels approach at larger angles. The deviations between the total fluxes at the channel exits from Table 3.2 hence originate exclusively from the forward-directed component of the neutron current causing the same effects. In the

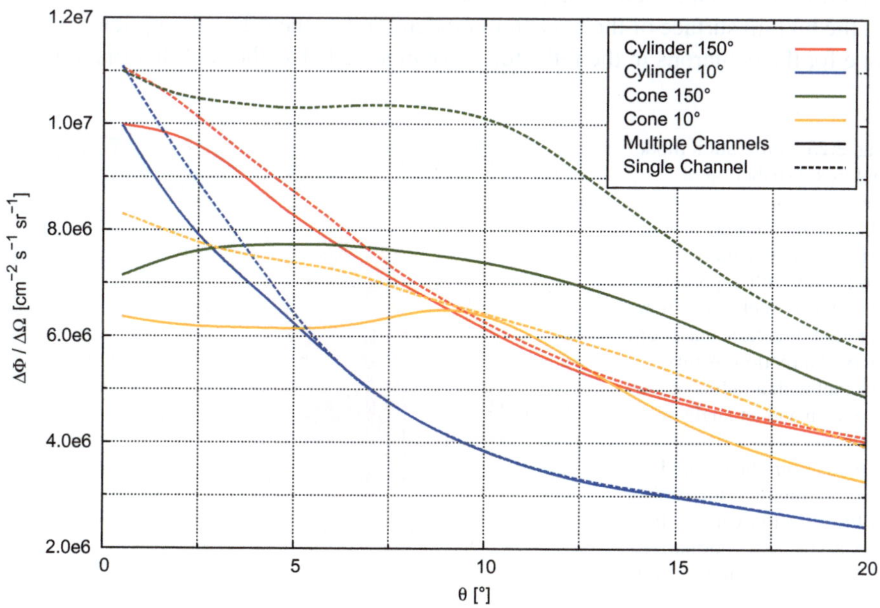

Fig. 3.22 Brilliance of thermal neutrons at the exit of a conical flux channel with $\varphi = 150°$. The radius at the channel exit is kept constant with $r = 1$ cm while the inner radius R is varied

case of multiple cylindrical flux channels with a small radius of 1 cm, the loss in peak brilliance (first data point at 0.5°) of 10% compared to a single channel is much lower than for conical channels with 37% (channel with $\varphi = 150°$) and 23% (channel with $\varphi = 10°$), respectively. The values for the peak brilliance determined up to $\theta = 1°$ are also summarized in Table 3.2.

It can be assumed that, in the case of small flux channels, the mutual impairment is in acceptable range even with more than four channels. However, in the following investigations, simulation models with single flux channels are used in order not to distort the results.

3.5 Impact of an Additional Reflector

According to Fig. 3.17, Sect. 3.4.1, the thermal flux at the channel exits of the Finger Moderator can be considerably increased by the use of a reflector. This section examines the impact of an additional reflector on the spectrum, the (peak-) brilliance and the pulse structure. In general, these are the key parameters when determining the instrument performance at a neutron source.

The simulation model used for the investigations is depicted in Fig. 3.23 and consists of a cylindrical beryllium moderator with $r = 31$ cm and $h = 41$ cm and a cylindrical graphite reflector of thickness 20 cm in radial direction and 29 cm in axial direction. The source definition presented in Sect. 3.2.3 is located at (0, 0, −0.1) cm on the bottom surface of the beryllium moderator. For the sake of simplicity, a beam tube for the deuterons in the reflector is not modeled. The thermal flux maximum is

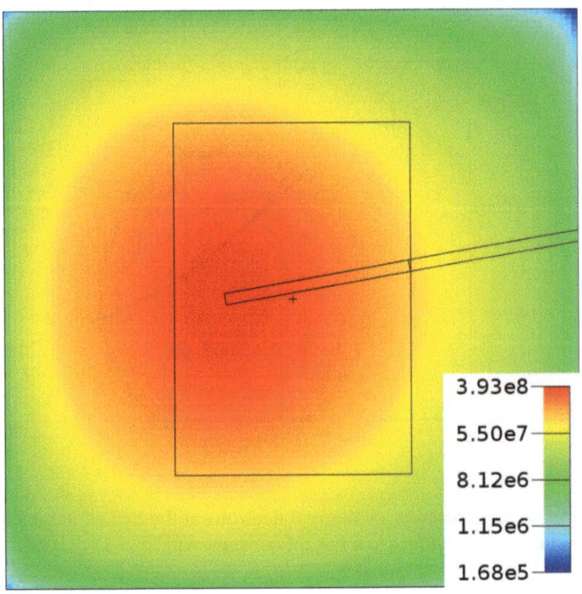

Fig. 3.23 Simulation model and flux distribution of thermal neutrons ($E \leq 30$ meV) of a cylindrical beryllium moderator with a graphite reflector of thickness 20 cm in radial direction and 29 cm in axial direction. A flux channel with a radius of $r = 1$ cm and an angle of incidence of $\varphi = 10°$ extends from the thermal flux maximum to the surface of the reflector. Unit of the flux is $cm^{-2}\,s^{-1}$ normalized to Trident operating at 10 Hz

located at a depth of 8.9 cm on the axis of the moderator. From this position, a flux channel with a radius of $r = 1$ cm and an angle of incidence of $\varphi = 10°$ extends to the surface of the reflector.

3.5.1 Spectrum and Brilliance

First, the influence of the reflector on the spectrum of the neutrons extracted from the moderator is examined. Only the forward-directed neutrons are considered for which the angle between direction of flight and axis of the flux channel is below 1°. For comparison, a simulation model without reflector as well as a model with reflector, but filled flux channel is simulated. Figure 3.24 shows the results.

The flux at 30 MeV can therefore be more than doubled by using a graphite reflector from $4.89 \times 10^{11}\,\mathrm{cm^{-2}\,s^{-1}}$ (blue curve) to $1.25 \times 10^{12}\,\mathrm{cm^{-2}\,s^{-1}}$ (red curve). However, the number of fast neutrons is also much higher. Between 10 MeV and 20 MeV, about four times more fast neutrons are extracted from the moderator.

As expected, the ratio of fast neutrons to thermal neutrons is shifted in favor of thermal neutrons for the bulk moderator-reflector assembly (green curve). However, only few thermal neutrons reach the surface of the bulk moderator so that the thermal

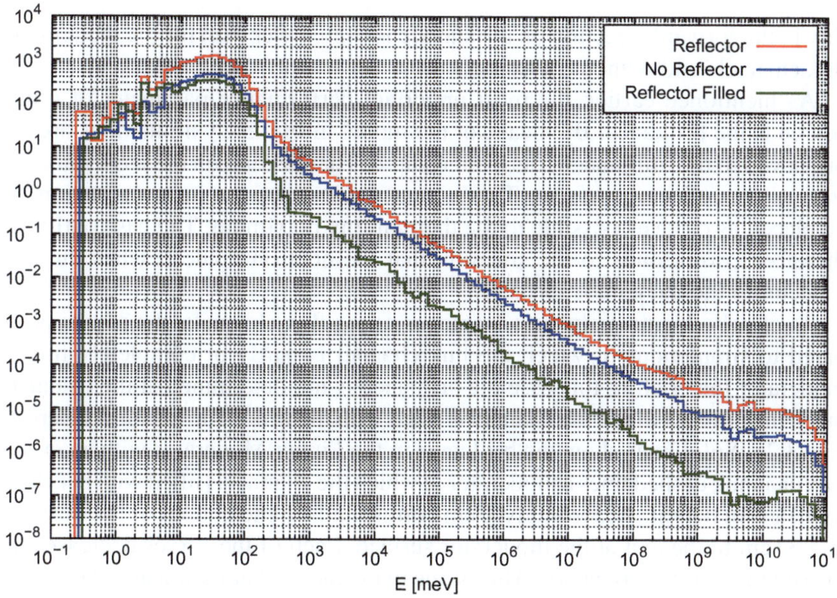

Fig. 3.24 Spectrum of neutrons leaving the moderator through the flux channel with angle of incidence of $\varphi = 10°$ in forward direction ($\theta \leq 1°$) for the assembly shown in Fig. 3.23, a bare beryllium moderator, and a bulk moderator-reflector assembly where the flux channel is filled with material

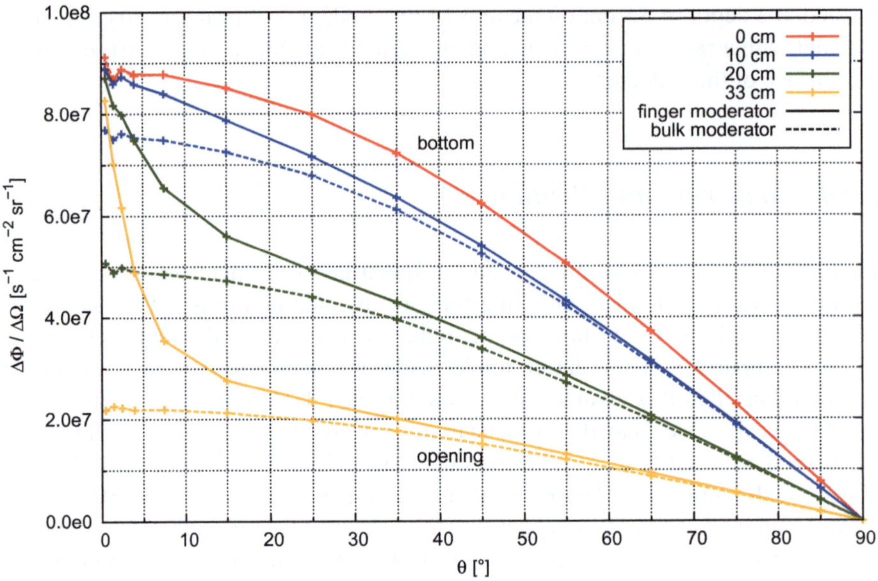

Fig. 3.25 Brilliance of thermal neutrons at different positions inside the flux channel with $\varphi = 10°$ (solid lines) and a bulk moderator, for which the channel is filled with material up to the measuring position (dashed lines)

flux in the forward direction on the surface is even lower than for the moderator with flux channels and without reflector.

As mentioned earlier, the most important performance parameter of a neutron source is the brilliance. By means of the model in Fig. 3.23 simulations are carried out in which the brilliance is determined at the channel exit on the moderator surface (not on the reflector surface). This is reasonable since the instrumentation should be placed as close as possible to the thermal flux maximum. The case outlined here therefore corresponds to the situation that the neutron guide reaches through the reflector to the moderator surface.

To clearly demonstrate the advantages of the flux channels, the results are compared to simulations in which the flux channel is partly or completely filled with moderator material. In this case the brilliance is not determined at the channel exit, but at the point to which the channel is filled with material. The results are depicted in Fig. 3.25. Solid lines show the brilliance in the Finger Moderator with an evacuated flux channel, dashed lines a moderator in which the channel is filled with beryllium up to the given height. Accordingly, 0 cm corresponds to the bottom of the flux channel and 33 cm to the exit at the moderator surface. Each of the curves decreases from a maximum at 0° to zero at 90°. This is due to the fact that emission at 90° corresponds to neutrons with flight direction tangential to the channel exit. For positions closer to the center of the moderator the decrease of the brilliance is rather flat whereas the brilliance at the channel exit rapidly decreases in the first few degrees and approaches the minimum with a flat tail. This behavior can be explained by the fact that tally surfaces closer to the center of the moderator also count neutrons which are emitted from the channel bottom and have a higher divergence, without being scattered by the inner walls of the flux channel. In the case of tally surfaces further outside the contribution to the brilliance for larger angles originates only from the outermost areas of the inner walls of the channels, where the neutron flux is significantly lower.

By comparing the maximum values at 0° it becomes evident, that neutrons which leave the moderator in forward direction mainly originate from the center of the moderator. This is not the case for bulk moderators in which the brilliance in forward direction is perturbed by the additional material. The peak brilliance is strongly decreasing for increasing depth inside the flux channel whereas for larger angles the brilliance approaches the values for the case of the Finger Moderator.

At this point it can be concluded that the brilliance in forward direction is extracted from the center of the moderator and is almost completely preserved by the flux channels.

3.5.2 Neutron Pulse Shapes

For the design of the experiments and in order to adapt the facility to the respective experiments in an optimum way, cf. Sect. 2.4, the temporal pulse structure of the generated neutrons is examined. The pulse structure is determined at the channel exits by means of a time-binned surface current tally. The energies are also divided into the regions cold, thermal, fast and total. For the flux channels with $\varphi = 90°$ and $\varphi = 150°$ the results are shown in the Figs. 3.26 and 3.27, respectively.

The initial pulse of fast neutrons below 300 ns correspond to the primary neutrons generated at Trident as can be seen in Fig. 3.3b and in Fig. 3.8c. The moderation time of the system amounts to 70 μs, in agreement with the theoretical value from Table 2.2, above which the pulse consists almost entirely of thermal neutrons and afterwards is slowly decreasing up to 1 ms. The background of overmoderated, cold neutrons is smaller by an order of magnitude. The flux of fast neutrons is even lower as indicated by the yellow curve.

The pulse shape at the 150° flux channel differs significantly in the height of the thermal pulse from the channel 90°. In contrast, the plateau of the thermal flux is less pronounced between 100 μs and 1 ms whereas the tail of the thermal pulses are of similar shape.

In order to estimate the background during the thermal pulses, the ratio of the thermal flux to the total flux as a function of the time are plotted for the channels with $\varphi = 10°, 40°, 90°$, and $150°$ in Fig. 3.28. As already mentioned, the thermal pulse reaches its full height after about 130 μs. The background of cold and fast neutrons amounts to 10% between 130 μs and 200 μs. Up to about 500 μs the background increases slightly, and remains constant at 13% lateron.

In a more extensive simulation the pulse structure is determined in forward direction in a cone with $\theta \leq 1°$ for a channel with $\varphi = 90°$. The results are depicted in

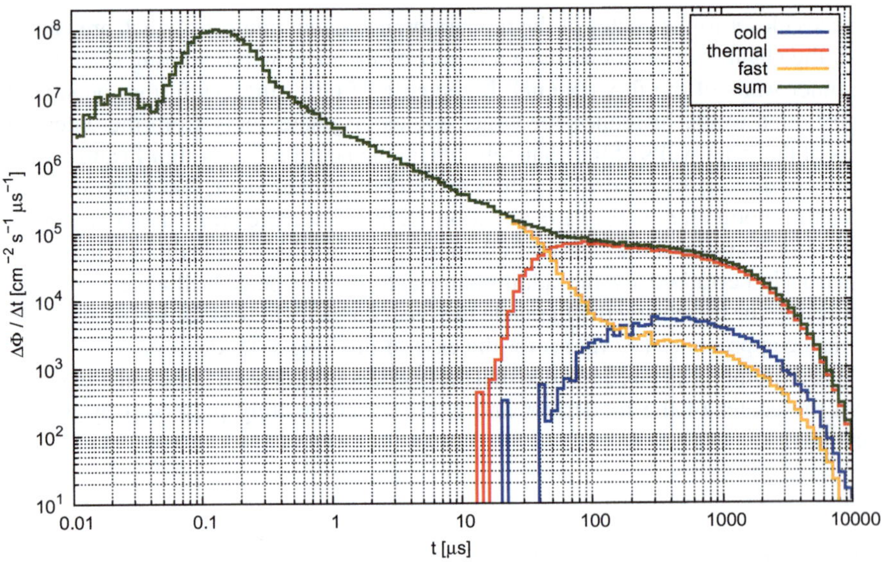

Fig. 3.26 Pulse shape of cold, thermal and fast neutrons at the exit of the channel $\varphi = 90°$. Values are normalized to Trident operating at 10 Hz

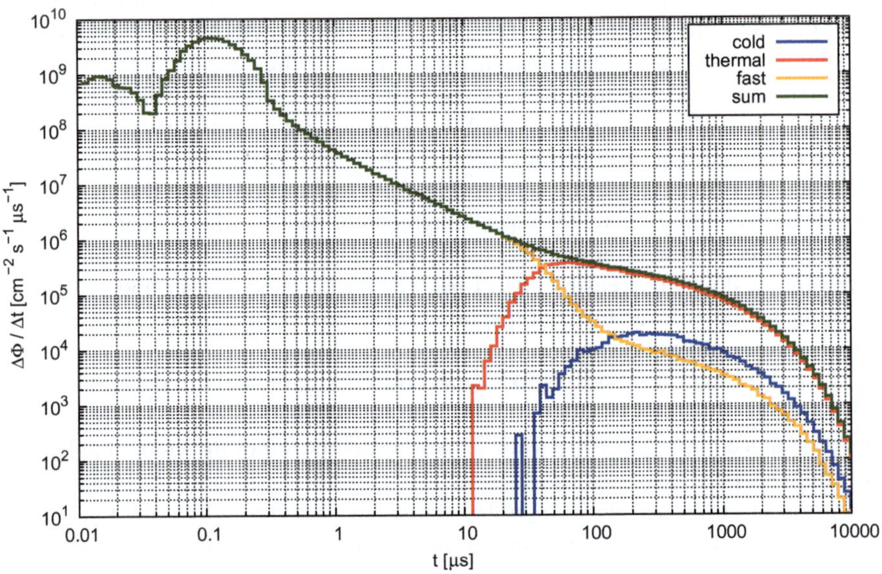

Fig. 3.27 Pulse shape of cold, thermal and fast neutrons at the exit of the channel $\varphi = 150°$. Values are normalized to Trident operating at 10 Hz

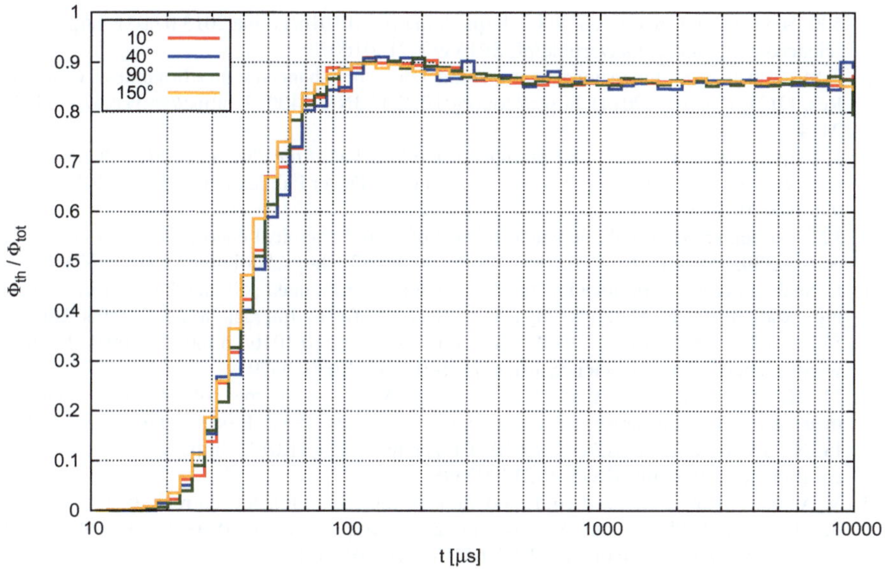

Fig. 3.28 Ratio of thermal flux to total flux at the channel exits in dependence to the time

Fig. B.3 in the appendix and indicate, that the neutron pulse in forward direction is slightly shorter, beginning at 100 μs and decreasing somewhat faster.

References

1. D. Jung et al., Characterization of a novel, short pulse laser-driven neutron source. Phys. Plasmas **20**(5), 056706, 9 (2013). https://doi.org/10.1063/1.4804640. https://doi.org/10.1063/1.4804640
2. M. Roth et al., Bright laser-driven neutron source based on the relativistic transparency of solids. Phys. Rev. Lett. **110**(4) (2013). https://doi.org/10.1103/PhysRevLett.110.044802. https://doi.org/10.1103/PhysRevLett.110.044802
3. D. Jung, Ion acceleration from relativistic laser nano-target interaction. Ph.D. thesis. Munchen: LMU Munich, Faculty of Physics, Jan. 6, 2012. https://edoc.ub.uni-muenchen.de/14074/1/Jung_Daniel.pdf
4. R.E. MacFarlane, A.C. Kahler, Methods for processing ENDF/B-VII with NJOY. Nucl. Data Sheets **111**(12), 2739–2890 (2010). ISSN: 0090-3752. https://doi.org/10.1016/j.nds.2010.11.001. http://www.sciencedirect.com/science/article/pii/S0090375210001006
5. R. Suzuki, GSYS2.4 manual. Second Edition-α. Technical Report Hokkaido University Hospital. Oct. 28, 2012. http://www.jcprg.org/gsys/2.4/gsys24-e.pdf
6. X-5 Monte Carlo Team. MCNP—A General N-Particle Transport Code, Version 5. LA-UR-03-1987. Technical report Los Alamos National Laboratory. Feb. 1, 2008. https://laws.lanl.gov/vhosts/mcnp.lanl.gov/pdf_files/la-ur-03-1987.pdf
7. K. Batkov, A. Takibayev, L. Zanini, F. Mezei, Unperturbed moderator brightness in pulsed neutron sources. Nucl. Instrum. Methods A: Accel Spectrom Detect Assoc Equip **729**,

500–505 (2013). ISSN: 0168- 9002. https://doi.org/10.1016/j.nima.2013.07.031. http://www.sciencedirect.com/science/article/pii/S0168900213010073

8. F. Mezei et al., Low dimensional neutron moderators for enhanced source brightness. J. Neutron Res. **17**(2), 101–105 (2014). https://doi.org/10.3233/JNR-140013. https://arxiv.org/pdf/1311.2474
9. J. Byrne, *Neutrons, Nuclei and Matter, An Exploration of the Physics of Slow Neutrons. An Exploration of the Physics of Slow Neutrons*. 2nd ed., Dover Publications, Dec. 14, 2011. ISBN: 978-0486482385
10. J.M. Carpenter, Thermally activated release of stored chemical energy in cryogenic media. Nature **330**, 358–360, Nov. 26, 1987. ISSN: 0028-0836. https://doi.org/10.1038/330358a0
11. E. Shabalin et al., Experimental study of swelling of irradiated solid methane during annealing. Nucl. Instrum. Methods Phys. Res. Sect. B: Beam Interact. Mater. Atoms **266**(24), 5126–5131 (2008). ISSN: 0168-583X. https://doi.org/10.1016/j.nimb.2008.09.022. http://www.sciencedirect.com/science/article/pii/S0168583X08010379
12. M. Huerta Parajon, E. Abad, F.J. Bermejo, A review of the cold neutron moderator materials: neutronic performance and radiation effects. Phys. Procedia **60**, 74–82 (2014). ISSN: 1875-3892. https://doi.org/10.1016/j.phpro.2014.11.012. http://www.sciencedirect.com/science/article/pii/S1875389214005598
13. V. Ananiev et al., Pelletized cold moderator of the IBR-2 reactor: current status and future development. J. Phys. Conf. Ser. **746**(1) (2016). https://doi.org/10.1088/1742-6596/746/1/012031. http://stacks.iop.org/1742-6596/746/i=1/a=012031
14. F. Cantargi, J.R. Granada, Thermal neutron cross-section libraries for aromatic hydrocarbons. Nucl. Instrum. Methods Phys. Res. Sect. B: Beam Interact. Mater. Atoms **268**(16), 2487–2491 (2010). ISSN: 0168-583X. https://doi.org/10.1016/j.nimb.2010.04.030. http://www.sciencedirect.com/science/article/pii/S0168583X10004209
15. U. Rucker, T. Bruckel, T. Cronert, J.P. Dabruck, R. Nabbi, Vorrichtung zur Erzeugung von thermischen Neutronenstrahlen mit Hoher Billanz und Herstellungsverfahren. PUB:(DE-HGF)23 Patent 16166567.4-1556 (EP3091540B1) (Germany). JCNS-2 / PGI-4 / JARA-FIT (2016). http://juser.fz-juelich.de/record/810621

Chapter 4
Optimization Studies on an Accelerator-Driven Neutron Source

Optimization studies for the accelerator-driven HBS are based on the simulation results on the laser-driven neutron source discussed in the previous chapter. Findings regarding the moderator material and the dependence of the neutron flux on the moderator configuration and in particular the idea of the Finger Moderator are therefore adopted to the accelerator-driven system. Another important aspect in the development of a concept for the accelerator-driven HBS is the target design, for which the neutron yield is investigated as a function of the incident particle type, its energy, and the target material.

In order to achieve the optimizations of the moderator-reflector assemblies more efficiently, a method is developed which uses the PTRAC feature of MCNP to produce a source file containing all relevant data of the phase space of the primary neutrons at the time they are produced in the target. Hence, the simulation of the initial ions is skipped in subsequent simulations by applying the source file as an input for the neutron transport calculations. Based on the resulting simulation model, the dependency of various performance quantities, such as spectrum, brilliance, and pulse structure, is investigated in order to determine the optimum configuration for each type of neutron scattering experiment.

In all simulations presented in this chapter the symmetry axis of the moderator is assumed to be in positive z-direction. Neutrons with energies from 0.568 meV (6.59 K, 12 Å) to 13.1 meV (152 K, 2.5 Å) are considered cold whereas neutrons in the energy range between 13.1 meV (152 K, 2.5 Å) and 127 meV (1474 K, 0.8 Å) are considered thermal. These energy ranges are listed in Table A.1 together with the corresponding neutron velocity, wavelength, and temperature.

J. P. Dabruck, *Target Station Optimization for the High-Brilliance Neutron Source HBS*, Springer Theses, https://doi.org/10.1007/978-3-030-05639-1_4

4.1 Target Development

4.1.1 The PTRAC Source File Approach

All possible reactions between a projectile (protons, deuterons, or electrons) and a target atomic nucleus in which neutrons are produced (cf. Sect. 2.2.1) have to be sampled with MCNP. In the desired energy regime for CANS the neutron yield is in the range of 1–10% (see Sect. 4.1.2). Thus, a large number of source particles do not contribute to the production of neutrons and a large amount of computing time is therefore wasted. Another even more important drawback of using protons or deuterons as initial particles is that MCNP cannot be executed multi-threaded using OMP. In this case parallelization is achieved only via MPI multiprocessing. However, with a high number of cores and mere MPI parallelization the speedup[1] is low as explained in Sect. 2.5.4.

For both these reasons, a method was developed in which the neutrons produced in the target can be used as source particles in MCNP without neglecting the underlying physics of the preceding projectile-target interaction. This method makes use of the `PTRAC` feature of MCNP and the possibility to implement user-defined source subroutines into the Fortran source code of MCNP. The idea is to create a source file in which all relevant quantities of the primary neutron's phase space are saved which are the position coordinates x, y, z, the cosines u, v, w of the angles of the flight direction with respect to the x-, y- and z-axes, the particle weight W, and the time t with respect to the source event of the corresponding initial ion. In addition to the advantages already mentioned, this file gives the opportunity to study the properties of the primary neutron in detail, see Sect. 4.1.2. The source file is read by the source subroutine in subsequent MCNP simulations where the stored parameters are assigned to the source particles. This approach is comparable to the MCNP built-in *surface source write* and *surface source read* routines [1]. In this approach particle parameters are recorded to a separate output file when the particle crosses a predefined surface. This output file can be used in subsequent runs substituting `SDEF` source definitions or fission sources in `KCODE` problems to increase the calculation performance. However, the built-in surface source capability exhibit some drawbacks as it cannot be used again in a single simulation if the number of starting particles `nps` is larger than the number of entries in the surface source file. Moreover, the surface source feature is limited to surfaces whereas the novel `PTRAC` source approach is independent of the geometry and additionally allows for an detailed analysis of the stored particles. If the end of the source file is reached but more histories are requested by `nps`, the subroutine will start again from the beginning. This results in multiple source particles having identical starting parameters. However, different random numbers are used to calculate the associated histories, which is the reason

[1] Speedup is a term used in computer science and mathematically describes the relationship between the serial and the parallel execution time of a program or algorithm.

for the paths of random walk of these histories being different and therefore being regarded as independent contributions to the tallies.

The procedure to generate the source file and the structure of the `SOURCE` subroutine are explained in detail in the Appendix C.

Validation The purpose of the validation of the `PTRAC` method is to ensure that the same results are obtained when using the source file as in simulations with protons as source particles. In the course of the validation the question is answered, how many entries in the source file are needed to sample the phase space, which is spanned by the parameters listed in the previous paragraph, of the primary neutrons with sufficient accuracy. For simplicity, a combined target moderator system is considered for the validation process. A beryllium cylinder of height 41.4 and 31 cm radius is bombarded with deuterons of energy 25 MeV emerging from the left in the center of its front surface. The spatial flux distribution of the created neutrons is determined with a resolution of 1 cm. The results are compared to simulations with source files of different sizes. Results are shown for the smallest and the full source file in Fig. 4.1. The deviation Δ of the flux distributions resulting from the source file to the simulation with deuterons is calculated according to formula 4.1 and is depicted on the left-hand side of each figure.

$$\Delta = \frac{\Phi_{\text{source file}} - \Phi_{\text{deuterons}}}{\Phi_{\text{deuterons}}} \tag{4.1}$$

A Gaussian error propagation of the relative errors of both simulations is performed according to Eq. 4.2 in order to exclude the statistical uncertainty of the simulations as a source of the deviation.

$$\sigma = \sqrt{\sigma(\Phi_{\text{deuterons}})^2 + \sigma(\Phi_{\text{source file}})^2} \tag{4.2}$$

Figure 4.1a shows the deviation of the flux distribution calculated with a source file with 10^4 entries to a "full" simulation with 2×10^9 deuterons as starting' particles which results in 3.2×10^7 primary neutrons. The maximum deviation amounts to -4%. The sharp fluctuations of the deviations in particular in the center of the cylinder are noticeable. As shown in Fig. 4.1b the relative total error in this area is 0%, so it can be concluded that the deviation is of systematic nature and the extent of the source file with 10^4 entries is too small to reproduce the correct ion-target interaction. For a source file with 10^5 entries, the maximum deviation amounts to -3%, but is about -1% in most areas (see Fig. B.7a in the Appendix). It is evident that this deviation of systematic nature since the overall relative error (cf. Fig. B.7b in the Appendix) is again $\sim$0%. Looking at Fig. B.7c it is apparent that a source file size with 10^6 starting particles is sufficient to reproduce the physics adequately; the deviation from the full simulation is 0% in most areas. Using the full source file with 2.6×10^6 entries brings no substantial improvement in the results (see Fig. 4.1). It can be concluded that for small volume targets a source file of size 10^6 is sufficient. If the method is applied to large models, e.g., in reactor simulations, a much larger source file has to be generated because of the much larger phase space.

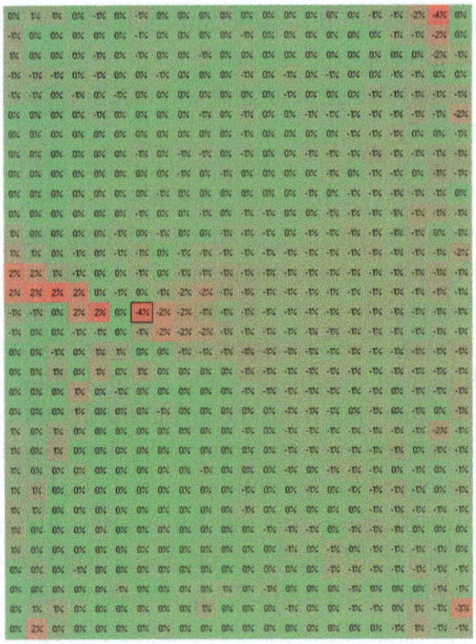

(a) Deviation using a source file with 10^4 particles. The largest value amounts to 4%.

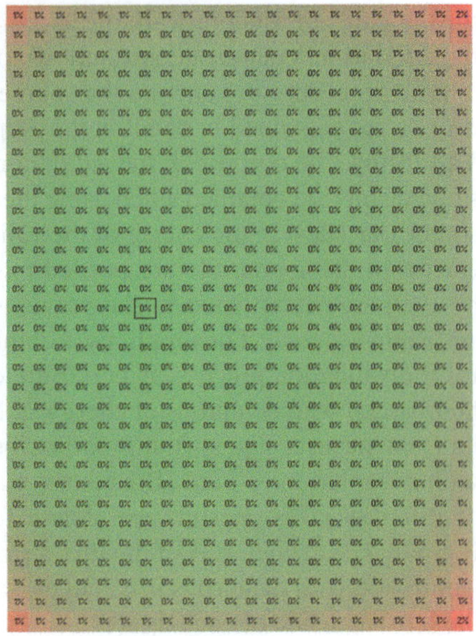

(b) Relative error ($\leq 2\%$) of the deviation according to equation 4.2.

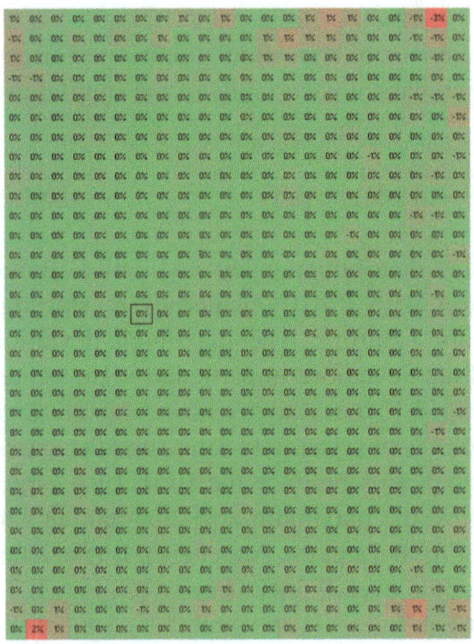

(c) Deviation using the full source file with 2.6×10^6 particles. The largest value amounts to 3%.

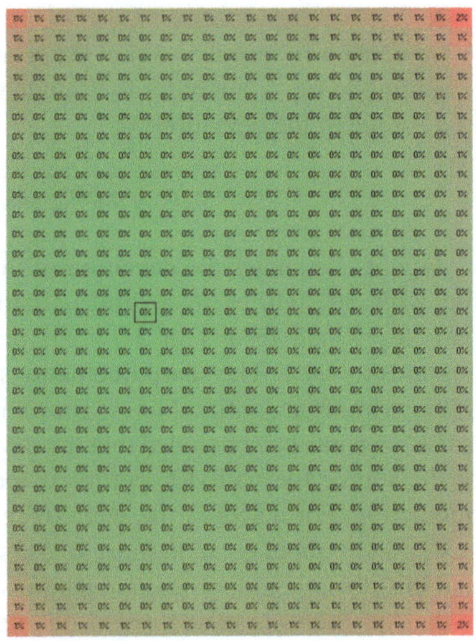

(d) Relative error ($\leq 2\%$) of the deviation according to equation 4.2.

Fig. 4.1 Deviation of the spatial neutron flux distributions calculated using source files of different sizes and a full simulation with deuterons as starting particles and its relative error

Table 4.1 Fit parameters and their standard deviation of benchmark simulations with a neutron source file and protons, respectively, for both simulation models

Only target	Source file	Protons
a [1/s]	$7.43 \times 10^4 \pm 96.91$ (0.13%)	19.25 ± 0.01 (0.05%)
b	$-4.82 \times 10^5 \pm 1.05 \times 10^5$ (21.75%)	-98.55 ± 18.27 (18.54%)
$t_{\text{init}}[s]$	6.49 ± 1.40 (21.62%)	5.12 ± 0.95 (18.49%)
With moderator	Source file	Protons
$a[1/s]$	2714.61 ± 2.55 (0.09%)	18.86 ± 0.01 (0.05%)
b	-32138.8 ± 23530 (73.22%)	-193.72 ± 138.30 (71.38%)
$t_{\text{init}}[s]$	11.84 ± 8.66 (73.13%)	10.27 ± 7.33 (71.33%)

In addition, the neutron spectrum is recorded throughout the volume with the help of a volume-based flux tally for all source files and the full deuteron simulation, see Fig. B.8 in the Appendix. The spectra match except some tiny deviations in the last two bins in the high energy regime.

Performance The gain in computation time by using a source file is determined for the target geometry, which is presented in Sect. 4.1.3, and protons as source particles.

A certain amount of time is gained for each history when using the source file because the proton transport is skipped. It must also be taken into account that only a small fraction of the protons generates neutrons at all. In addition, the read-in time of the source file can have a negative impact on the performance, depending on its size. All these effects mean that the speedup by using the source file strongly dependent on the geometry. For this reason, two benchmark studies are conducted, one study without moderator and one with a cylindrical beryllium moderator ($r = 31$ cm, $h = 41$ cm).

All calculations are performed on an Intel Xeon E5-2680 with 2.7 GHz and 2 GB RAM single-threaded with only one process so that the communication overhead of MPI and OMP does not distort the results. In both cases (source file with 2×10^6 entries and protons) several simulations are performed with different numbers of starting particles `nps`. Since only the generated primary neutrons are relevant, the `nps` values of the protons must be multiplied by $0.05897\, {}^{n}/_{p}$ according to Table 4.2. A linear fit of the form

$$n_{\text{prim}}(t) = b + a\,t$$

is performed where a denotes the number of primary neutrons per second. The initialization time of the simulations is given by $t_{\text{init}} = -{}^{b}/_{a}$ whereby the read-in time of the source file is determined by subtracting the initialization time of the proton simulation from the initialization time of the source file simulation. Results are depicted in Fig. 4.2 and listed in Table 4.1.

For the configuration without the moderator 7.4×10^4 primary neutrons can be calculated per second using the source file, while only 19.1 primary neutrons are calculated per second when protons are used as source particles. The speedup for

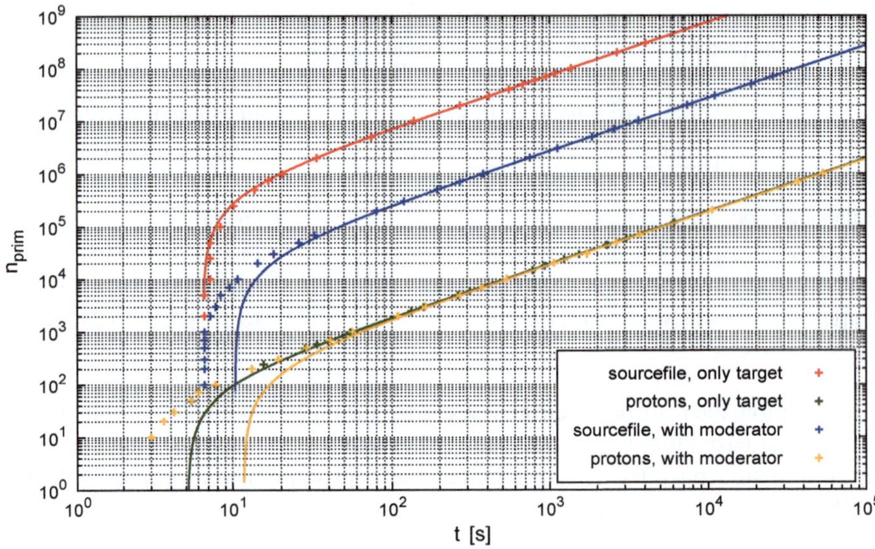

Fig. 4.2 Number of histories depending on computation time (one process, single threaded) for protons ($E = 50\,\mathrm{MeV}$) as source particles and usage of a corresponding source file (red) and linear fit for both configurations

this configuration is therefore 3860. The read-in time of the source file amounts to 1.37 s. For the configuration with moderator the speedup amounts to 144 and the read-in time of the source file to 1.57 s. The read-in times of the source file for both configurations show good agreement. However, it must be taken into account, that for simulations with protons as initial particles additional cross section libraries are loaded by MCNP. The true value of the read-in time for the source file is thus even a bit lower.

It is noticeable from Fig. 4.2 that the curves for the two configurations using protons are nearly identical. Therefore, it is evident that the fraction of neutrons in the mean total runtime is relatively low compared to the protons even when using low-absorbing beryllium (cf. Sect. 2.3.1). It must also be noted that only in 5.8% of the histories neutrons are produced at all. From this it can be concluded that the speedup will be significantly larger by using a source file for simulation models with even lower proton energy.

4.1.2 Projectiles and Target Materials

As mentioned in Sect. 4.1.1, the source files generated with the PTRAC method can be applied to study the properties of the primary neutrons, which are the neutron yield per initial ion, their spectrum, their angular distribution and brilliance, respectively, the angularly resolved spectrum, and the spatial distribution. Since single-event data is stored in the PTRAC files, a binning has to be performed to determine the corresponding frequency distributions. For this purpose a JAVA program is developed by [2] that can bin the data in two dimensions, for example energy E and flight direction u, v or w. The binning tool is used whenever a frequency distribution is created by means of particle data derived with `PTRAC`. A detailed explanation of the source code is found in Appendix C.2.

4.1.2.1 Neutron Yields for Different Ions and Targets

Source files are generated for different combinations of ions, targets and incident ion energies for which the associated neutron yields can be determined. This approach is independent of the target geometry (as long as the target is larger than the penetration depth of the ions), since the primary neutrons are tracked directly at the source and not, for example, by a tally on the target surf ace.

Results are listed in Table 4.2. For protons on beryllium the simulations are performed using the ENDF/B-VII.1 database and in further simulations with the TENDL-2014 database, see Sect. 2.5.4. It is evident that a large discrepancy occurs between the two databases. Using TENDL data neutron production is underestimated by a factor of 3 for energies of 50 MeV and below. For higher energies the discrepancy is decreasing. A comparison with experimental data demonstrates that the ENDF data reproduces the neutron yield with good agreement. This is further discussed later on. Further simulations on the neutron yield from protons using also TENDL-2015 data are conducted by [3] and depicted in Fig. B.4 in the Appendix.

For deuterons only TENDL data is available. However, the resulting neutron yield is close to the neutron yield from protons calculated with ENDF. From a physical point of view this is questionable because deuterons bring their own neutron so that additional reaction channels with neutrons in the final state are possible as discussed in Sect. 2.2.1. It is therefore assumed that the TENDL data underestimates the neutron production not only for protons but also for deuterons. To elucidate this issue experiments with protons and deuterons at the ring accelerator COSY [4] or others are planned. Figure 4.3 shows the yields of protons and deuterons impinging on beryllium for incident energies between 5 and 100 MeV. Additionally the empirical formula based on the measurements of [5] are plotted for the investigated energy range (23– 80 MeV). As the energy-dependent neutron yield calculated using ENDF is in good agreement with the experimental results, this dataset is used for all further simulations with protons if not stated otherwise.

Lithium is also being investigated as a target material, as it offers two major advantages over beryllium. First, the spectrum of neutrons produced is remarkably softer for low proton energies, as can be seen in Fig. 4.4. In addition, the target can be designed with liquid lithium. Thus the deposited heat load can be dissipated significantly better [6–8]. However, the low overall neutron yield should only be accepted if thermal or epithermal neutrons with ultra-short pulses are strictly required by the experiments.

An alternative to neutron production with ions and light metals is the bombardment of heavy metal targets with electrons. When high-energy electrons pass through high-

Table 4.2 Neutron yields of different ion-target combinations depending on the energy of the incoming ions calculated with TENDL-2014 and ENDF/B-VII.1, respectively. The data for electrons is taken from [10]

	E[MeV]	Dataset	n/n_{ion} (%)	n/s/μA	n/s/kW
p → Be	5	ENDF	0.049	3.04×10^{9}	6.08×10^{11}
		TENDL	0.015	9.49×10^{8}	1.90×10^{11}
	10	ENDF	0.339	2.12×10^{10}	2.12×10^{12}
		TENDL	0.110	6.86×10^{9}	6.86×10^{11}
	25	ENDF	2.195	1.37×10^{11}	5.48×10^{12}
		TENDL	0.638	3.98×10^{10}	1.59×10^{12}
	50	ENDF	6.211	3.88×10^{11}	7.75×10^{12}
		TENDL	2.597	1.62×10^{11}	3.24×10^{12}
	80	ENDF	10.004	6.24×10^{11}	7.81×10^{12}
		TENDL	6.537	4.08×10^{11}	5.10×10^{12}
	100	ENDF	12.533	7.82×10^{11}	7.82×10^{12}
		TENDL	9.791	6.11×10^{11}	6.11×10^{12}
d → Be	5	TENDL	0.063	3.94×10^{9}	7.89×10^{11}
	10	TENDL	0.265	1.65×10^{10}	1.65×10^{12}
	25	TENDL	1.608	1.00×10^{11}	4.02×10^{12}
	50	TENDL	5.724	3.57×10^{11}	7.15×10^{12}
	100	TENDL	19.602	1.22×10^{12}	1.22×10^{13}
p → Li	2	TENDL	0.002	1.02×10^{8}	5.11×10^{10}
	5	TENDL	0.050	3.15×10^{9}	6.30×10^{11}
	10	TENDL	0.164	1.02×10^{10}	1.03×10^{12}
	25	TENDL	0.471	2.94×10^{10}	1.18×10^{12}
e → U	20		0.396	2.47×10^{10}	1.24×10^{12}
	50		2.520	1.57×10^{11}	3.15×10^{12}
	100		5.695	3.55×10^{11}	3.56×10^{12}
	150		8.710	5.44×10^{11}	3.62×10^{12}
	200		11.861	7.40×10^{11}	3.70×10^{12}
e → Ta	50		1.400	8.74×10^{10}	1.75×10^{12}
e → W	50		1.500	9.36×10^{10}	1.87×10^{12}

Z materials, intense Bremsstrahlung is produced, which in turn undergo photonuclear interactions resulting in the generation of neutrons. Investigations are presented in [9, 10] with the results summarized in Table 4.2. At high energies the neutron yield is comparable to the ion-beryllium interaction. A disadvantage is that the neutron yield increases only linearly with the energy, so that the number of neutrons produced per accelerator power is almost constant above 50 MeV. Another serious drawback is the very high gamma background. Not only for reasons of radiation protection, but also especially for neutron scattering experiments, a comprehensive shielding is required which counteracts the low purchase costs of the accelerator. The simulation results

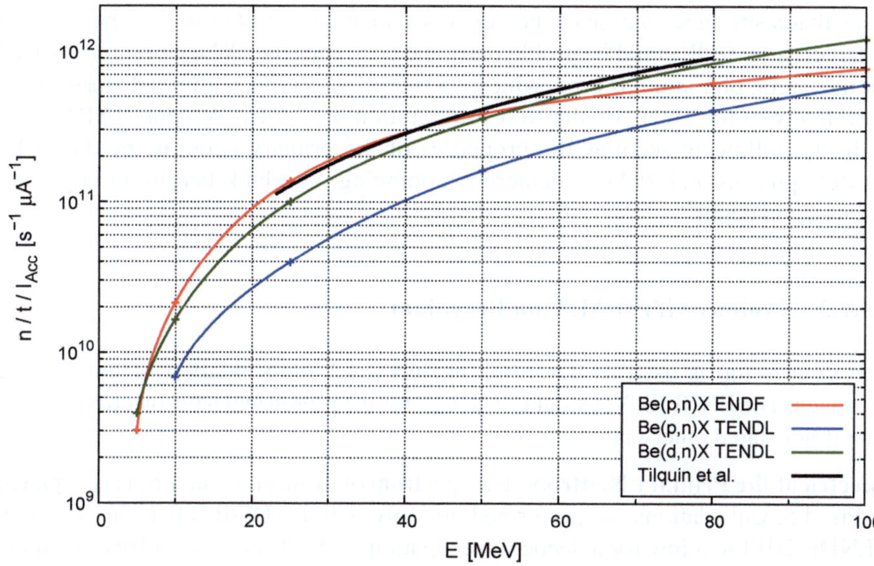

Fig. 4.3 Neutron yields of protons and deuterons on beryllium in dependence to the ion energy. Calculations are performed with ENDF/B-VII.1 for protons and with TENDL-2014 for protons and deuterons. An empirical curve derived from measurements of Tilquin et al. is also included [5]

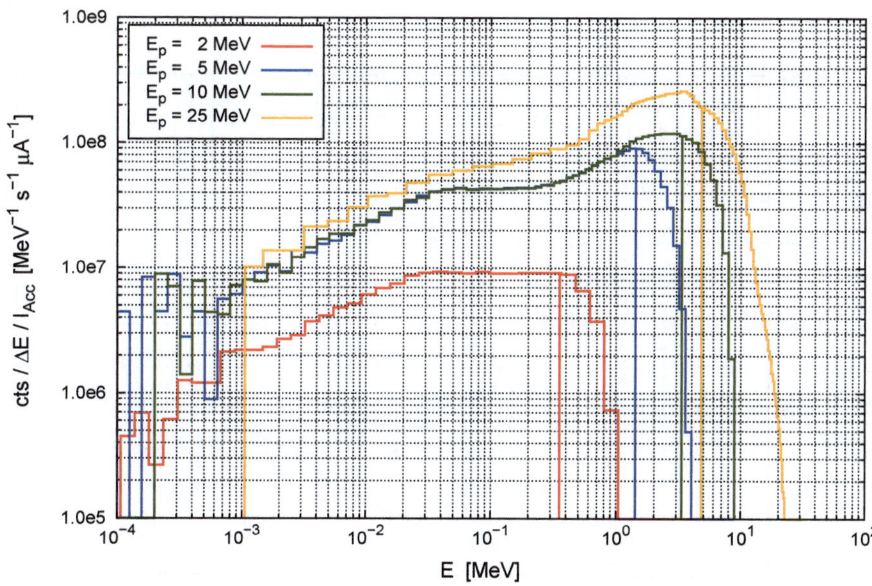

Fig. 4.4 Spectra of primary neutrons for protons with different energies impinging on lithium. Mean energies are 0.35, 1.43, 3.31, and 4.77 MeV for increasing proton energy

show that a strong activation of the target is also to be expected so that the disposal of targets after EOL could be problematic and cost-intensive. When uranium is used for the target, plutonium is bred, which leads to considerable additional work on the compliance to the Treaty on the Non-Proliferation of Nuclear Weapons (NPT).

In the following sections the properties of the primary neutrons produced by 50 MeV protons and 25 MeV deuterons impinging on a bulk beryllium target are discussed in detail.

4.1.2.2 Protons with 50 MeV on Beryllium

The same setup is used as described in Sect. 4.1.1 for validation of the PTRAC approach. The ENDF/B-VII.1 dataset is used for the proton and neutron cross section data if not stated otherwise.

Spectra of the Primary Neutrons The spectrum of the primary neutrons is depicted in Fig. 4.5. Calculations are performed not only with ENDF/B-VII.1 but also with TENDL-2014 to allow for a deeper investigation of the discrepancy observed in the neutron yields, cf. Table 4.2.

According to the results the spectrum is soft compared to the spectrum produced by laser-induced ion acceleration (cf. Fig. 3.3a) having a mean energy of 11 MeV (ENDF/B-VII.1). Energies above 50 MeV are kinematically forbidden. The max-

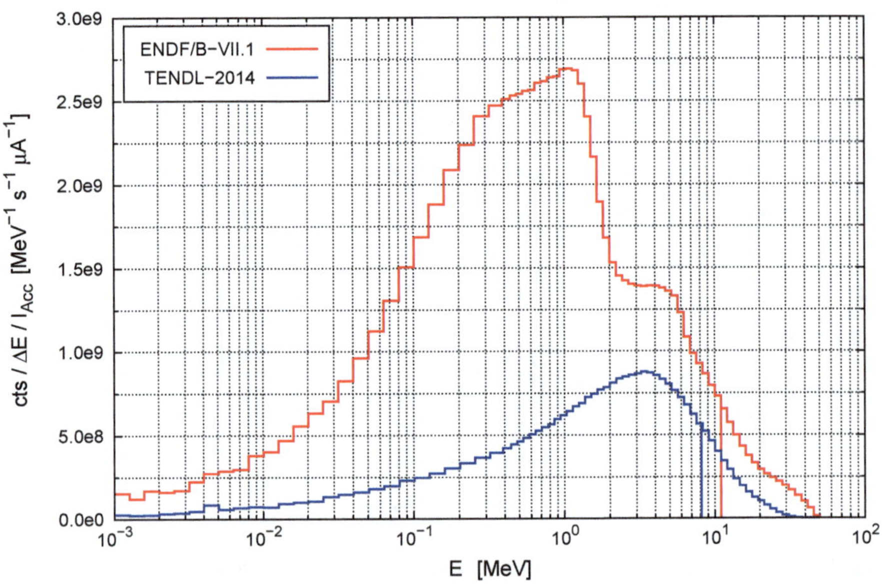

Fig. 4.5 Primary neutron spectrum calculated with ENDF/B-VII.1 and TENDL-2014 data sets. The mean energy amounts to 10.97 and 8.11 MeV, respectively

imum flux is reached at 1.1 MeV. Between 2.5 and 5 MeV the spectrum exhibits a plateau which is half as high as the maximum which presumably arises from additional reaction channels being established above a specific threshold energy. In case of using TENDL, the spectrum is significantly lower, with the total number of neutrons produced being a factor of 2.4 below that of ENDF-based calculations. Interestingly, the maximum of the spectrum occurs at 3.3 MeV; at the same location of the plateau of the ENDF spectrum. For decreasing energy, the spectrum also decreases strictly monotonously, in contrast to the spectrum produced using ENDF. This suggests that in the TENDL data the reaction channels leading to the production of neutrons with lower energies around the maximum of the ENDF spectrum are not taken into account.

Angular Distribution of the Primary Neutrons The angular distribution of the primary neutrons is depicted in Fig. 4.6. Calculations are performed with ENDF and TENDL. The angular distribution is measured with respect to the flight direction of the incident proton beam.

Neutron-producing reactions of protons are compound nucleus reactions where the reaction products have no predominant direction. However, this only holds for the center of mass system of the incident proton and the target nucleus at rest. For this reason the primary neutrons are strongly forward-directed [11–13]. The angular distribution in the case of using TENDL, however, is much flatter. The ratio of the forward-directed neutrons to the rear-facing neutrons amounts to 10.0 for ENDF and

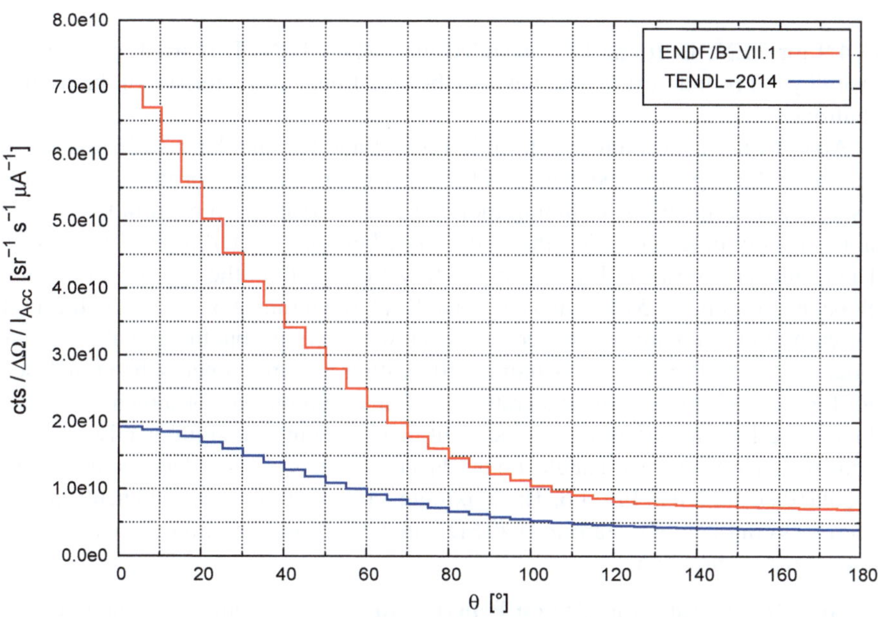

Fig. 4.6 Angular distribution of primary neutrons with respect to the incoming proton beam calculated with ENDF/B-VII.1 and TENDL-2014 data sets

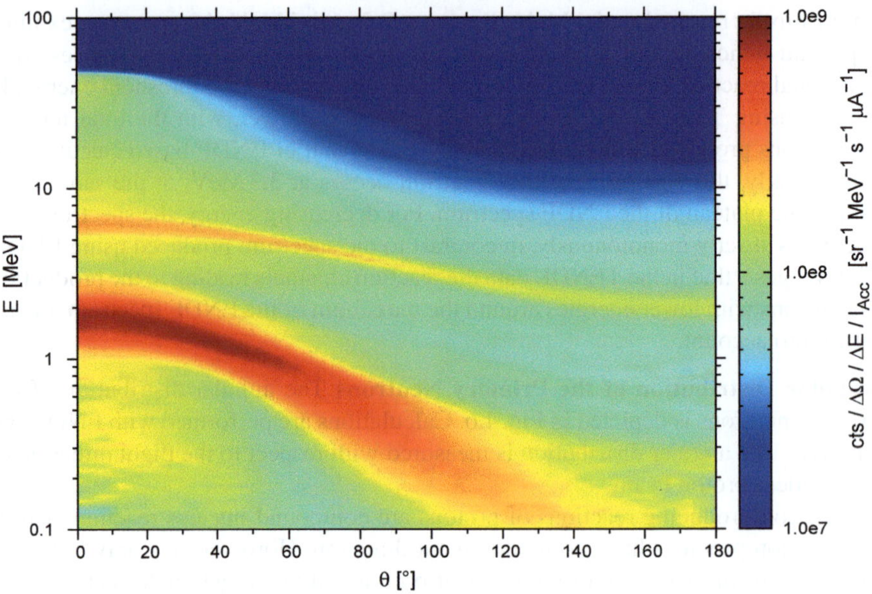

Fig. 4.7 Heat map of the primary neutron spectrum in dependence to the emission angle with respect to the proton beam. This is a logarithmic plot to emphasize the two maxima which are also visible in the integrated spectrum in Fig. 4.5. The darkest blue tone corresponds to a value of 0

to 4.9 for TENDL. It can be concluded that, when TENDL is used, not only the overall yield is underestimated, but also the directional characteristic of the primary neutrons.

Apart from the absolute spectrum in Fig. 4.5 and the angular distribution shown here, also the angular dependence of the spectrum is relevant for the target design. Using the aforementioned binning tool, the PTRAC data are binned in two dimensions, namely in E and w (cosine of the angle between flight direction and z-axis). The results are shown in Fig. 4.7. In the forward direction, the spectrum exhibits a global maximum at 1.5 MeV and a second, local maximum at 6.1 MeV. As the angle increases, both maxima are shifted towards lower energies, but the global maximum much stronger. The local maximum shifts within a narrow energy band and does not form a maximum in the integrated spectrum but a plateau as shown in Fig. 4.5. Taking the logarithmic scale into account, it is evident that significantly less neutrons with a high emission angle are produced, which is in accordance with the total angular distribution in Fig. 4.6. The maximum energy of the generated neutrons also decreases continuously, starting at 49 MeV in the forward direction down to 11 MeV in the reverse direction.

Spatial Distribution and Proton Penetration As a consequence of the ion deposition in the target accumulation of hydrogen and hence embrittlement of the target material, cf. Sect. 2.2.2, takes place. The PTRAC feature also allows to investigate

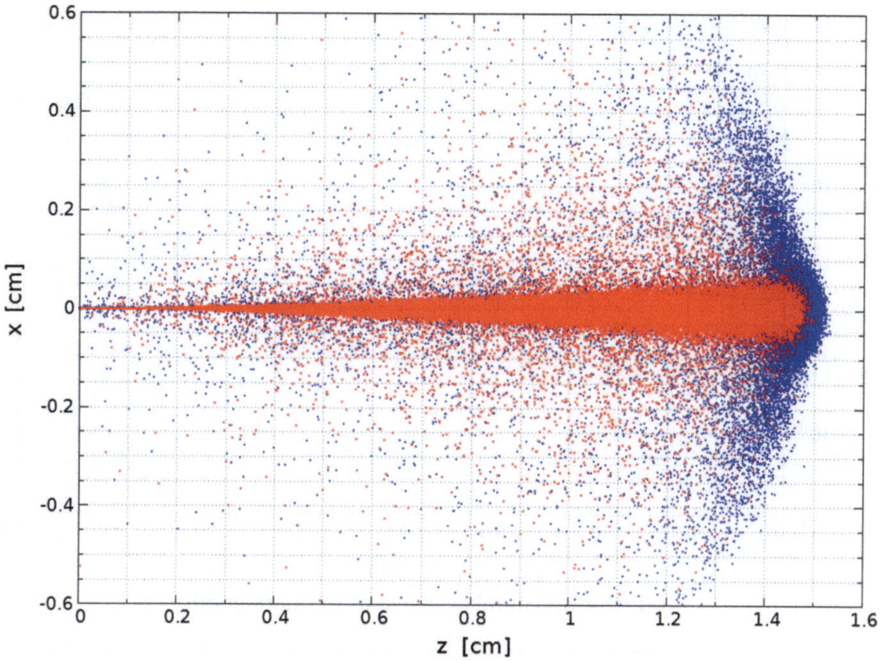

Fig. 4.8 Scatterplot of neutron production (red) and proton termination (blue) events in beryllium projected onto the z-x-plane for an initial proton energy of 50 MeV

the ion deposition. For this purpose, the event type to be tracked is set to `ter` (termination). Further filtering is not required. The definition of the `PTRAC` card is given in Appendix C.

A first insight into the results is given in the form of the scatter plot in Fig. 4.8. It shows the positions of individual proton termination events and additionally the birthplaces of the primary neutrons in the x-z plane. Due to the rotational symmetry of the problem, the y-direction can be neglected. The maximum penetration depth of the protons amounts to 15.2 mm which is in good agreement with calculation results obtained with SRIM-2013.00 (14.6 mm). The pencil beam of protons is broadened due to elastic scattering in the material. Most events occur in a cylinder with radius 0.6 mm on the z-axis.

Based on the scatter plot of the single-event data no information about the frequency distribution of the positions is obtained. By applying a binning the number of protons deposited per line segment and the number of neutrons produced can be determined as given by Fig. 4.9. Based on the characteristic shape of the stopping power ($-\mathrm{d}E/\mathrm{d}z$) of protons in beryllium a distinct Bragg peak can be seen at 1.47 cm. Remarkably, the neutron production profile is very flat along the entire path of the protons up to their Bragg peak. If the data is presented cumulative and normalized to 100%, as shown in Fig. 4.9b, the impact on the neutron yield of changing the target thickness can be very well estimated. It is evident that, for example, at 1.4 cm

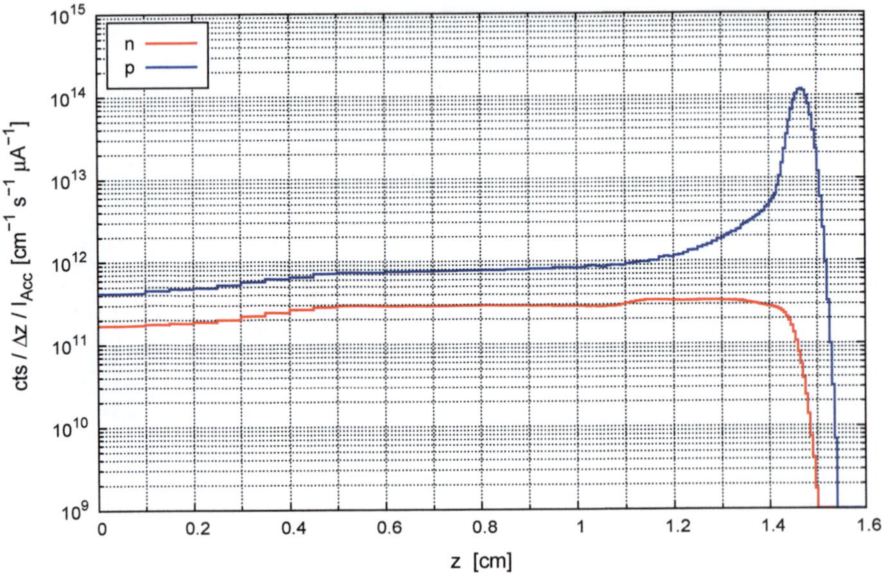

(a) Absolute number of events per unit length.

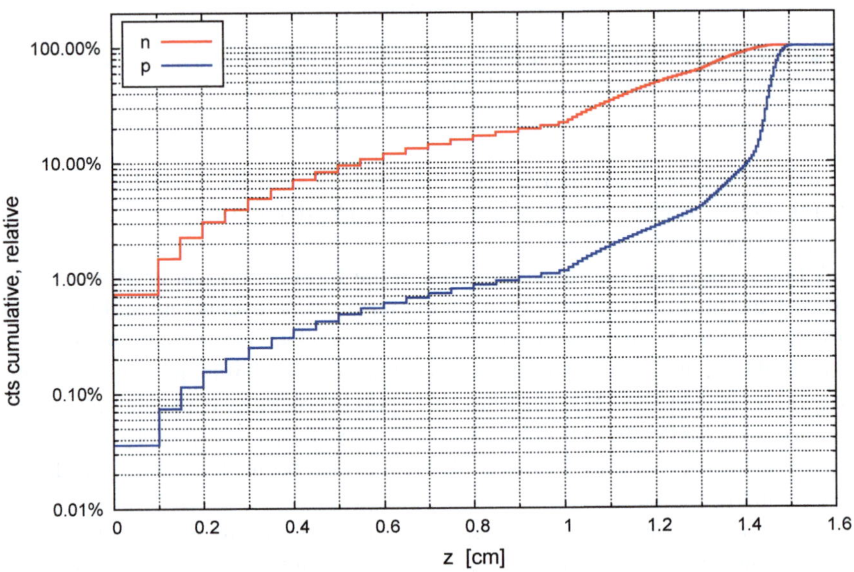

(b) Cumulative relative representation.

Fig. 4.9 Neutron production and proton termination in dependence to the depth inside a beryllium bulk target for 50 MeV protons

of target thickness 10% neutrons are lost, while only 9% of protons are deposited. With respect to hydrogen embrittlement due to proton deposition, the life time of the target can be increased significantly by limiting the target thickness [14–17], see Sect. 2.2.2.

4.1.2.3 Deuterons with 25 MeV on Beryllium

The same setup is used as described in Sect. 4.1.1 for validation of the PTRAC approach. The simulations are performed using MCUNED in combination with the TENDL-2014 data.

Spectra of the Primary Neutrons The spectrum of the primary neutrons produced by 25 MeV deuterons is depicted in Fig. 4.10. Calculations are performed not only with MCUNED but also with MCNP6.1. The results are in very good agreement.

The maximum value occurs at an energy of 3.2 MeV while the mean energy amounts to 5.9 MeV. In contrast to the primary neutron spectrum when using ENDF for 50 MeV protons, the shape of the spectrum is very smooth, which may be due to the fact that the interaction between the deuteron and the nucleus in the TENDL data is strongly idealized. It is conceivable that the real spectrum may also contain additional local maxima, which arise from additional reaction channels.

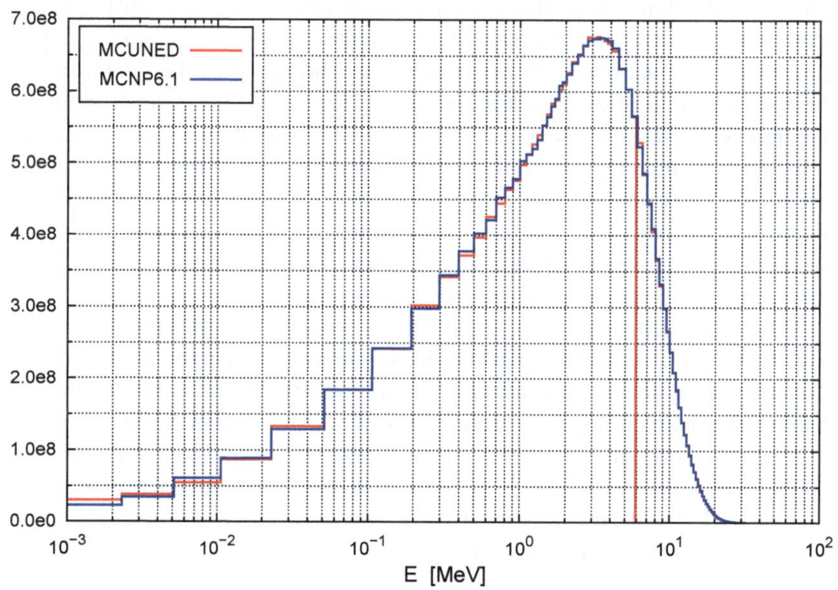

Fig. 4.10 Spectrum of primary neutrons normalized to one initial deuteron. Values are calculated with MCNPx + MCUNED and MCNP6.1, both with the TENDL-2014 database. The mean energy amounts to 5.89 MeV

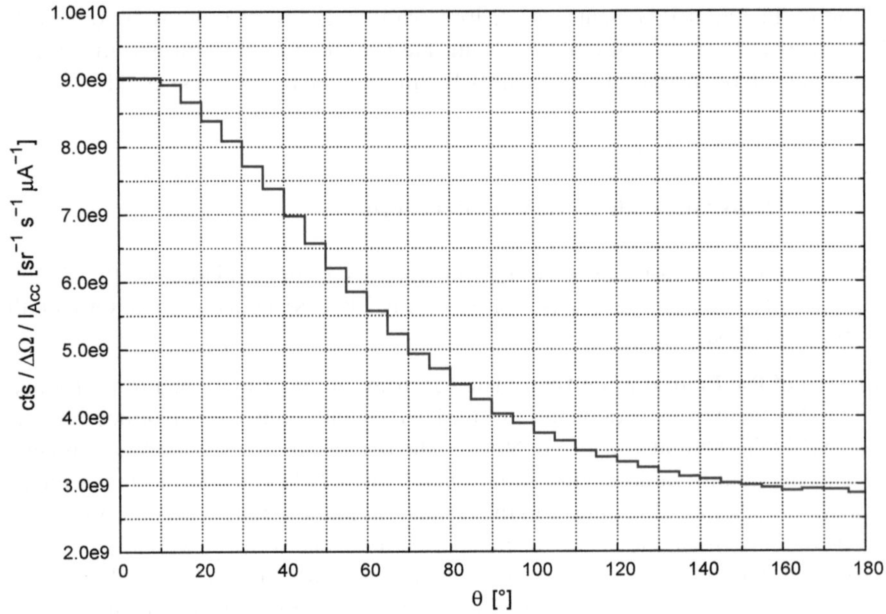

Fig. 4.11 Angular distribution of primary neutrons with respect to the incoming deuteron beam

Angular Distribution of the Primary Neutrons The angular distribution of the primary neutrons is depicted in Fig. 4.11. In addition to the compound-core reactions, direct nuclear reactions such as stripping and break-up of the deuterons occur in the coulomb field of the target atom nuclei. These reactions are strongly forward-directed as discussed in Sect. 2.2. For the angular distribution shown here the ratio of the forward-directed neutrons to the backwardly emitted neutrons amounts to 3.2. Despite the additional reaction channels, in which the neutrons are preferably emitted forwardly, this ratio is significantly lower than for 50 MeV protons, see Fig. 4.6. As for protons, however, it is also assumed that TENDL underestimates the directional characteristics of the primary neutrons also for deuterons.

The angular dependence of the spectrum is shown as a surface plot in Fig. 4.12a and as a heat map in Fig. 4.12b. Since the shape is less steep than for 50 MeV protons, the plots are shown linearly in favor of better readability. The maximum of the spectrum in the forward direction occurs at an Energy of 5.8 MeV and in the backward direction at 2.5 MeV. Thus, the dependency of the maximum flux value to the emission angle is relatively weak. As for the 50 MeV protons, the maximum energy of the neutrons emitted backwardly amounts to 10 MeV and in the forward direction to 25 MeV, which is the maximum energy kinematically allowed.

Spatial Distribution and Deuteron Penetration The scatterplot can be seen in Fig. 4.13. The maximum penetration depth of the deuterons amounts to 2.5 mm which is in good agreement with calculation results obtained with SRIM-2013.00 (2.4 mm). Most events occur in a cylinder with radius 0.25 mm on the z-axis.

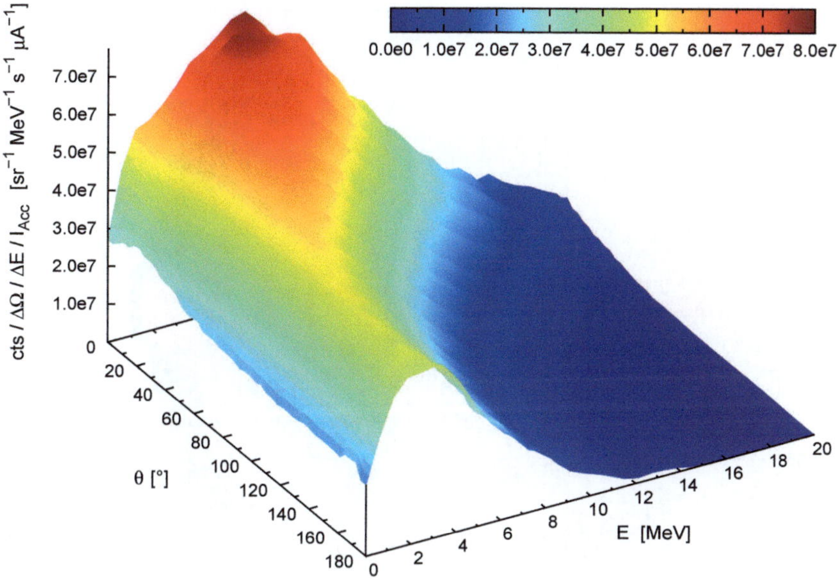

(a) Surface plot.

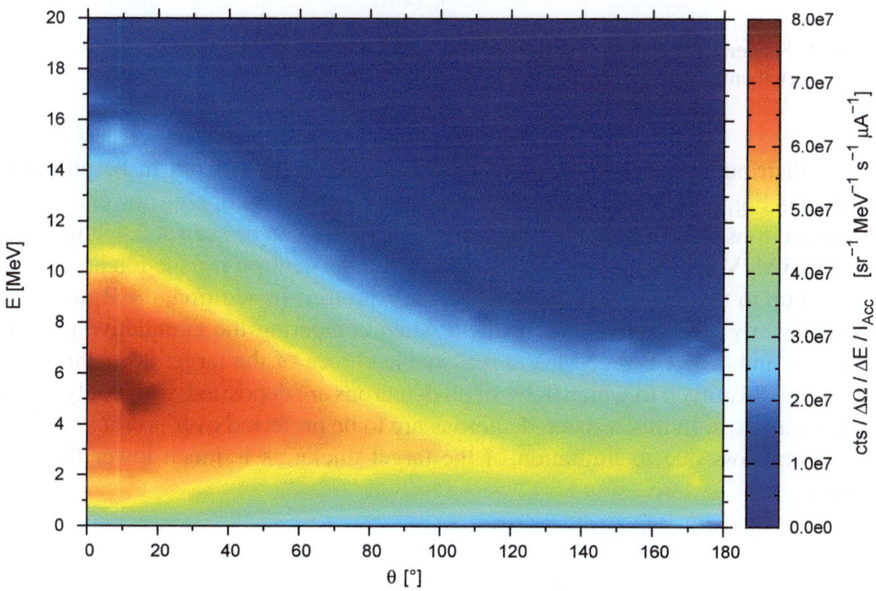

(b) Heatmap.

Fig. 4.12 Spectrum of primary neutrons in dependence to the emission angle with respect to the deuteron beam

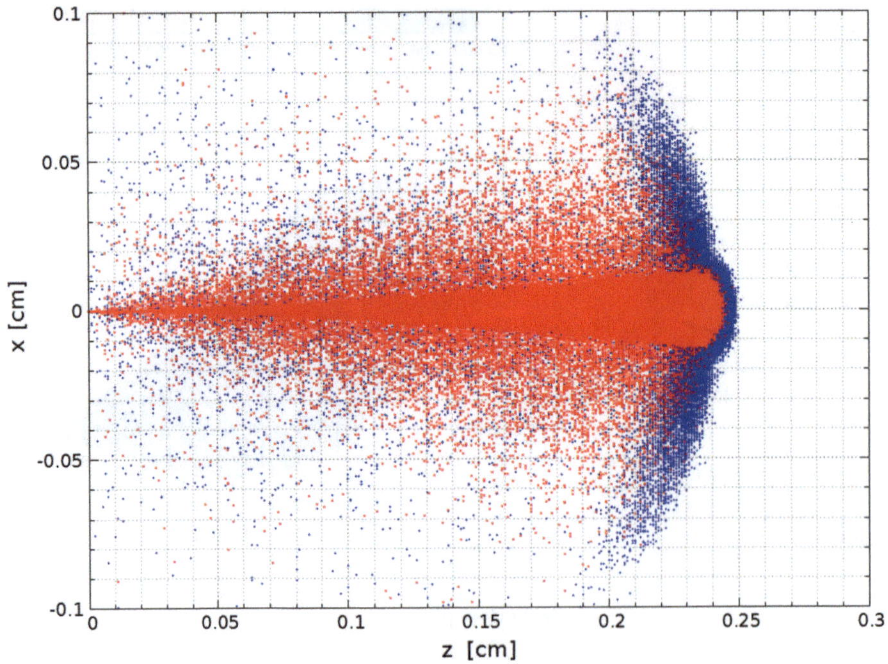

Fig. 4.13 Scatterplot of neutron production (red) and deuteron termination (blue) events in beryllium projected onto the z-x-plane for an initial deuteron energy of 25 MeV

The distribution of events depending on the penetration depth in beryllium is depicted in Fig. 4.14. Based on the characteristic shape of the stopping power ($-{}^{\mathrm{d}E}/_{\mathrm{d}z}$) of charged ions in bulk materials a distinct Bragg peak can be seen at 2.4 mm. In the case of 25 MeV deuterons, the ion deposition and the neutron production are almost constant up to the Bragg peak than for 50 MeV protons In addition, the Bragg peak is significantly narrower. This is evident when comparing the cumulative plots in Figs. 4.9b and 4.14b. For 25 MeV a reasonable choice of the target layer thickness would be 2.25 mm up to which 0.2% of all deuterons are deposited, while only 2% of neutrons are lost. In this respect, deuterons are to be preferred over protons because the neutron loss due to limitation of the target thickness is much lower than for protons.

4.1.3 *Target Shape and Geometry*

The following sections describes the approach and the results aiming at designing the target in detail. These include the geometric shape of the target as well as the selection of materials for the irradiated layer and the backing plate.

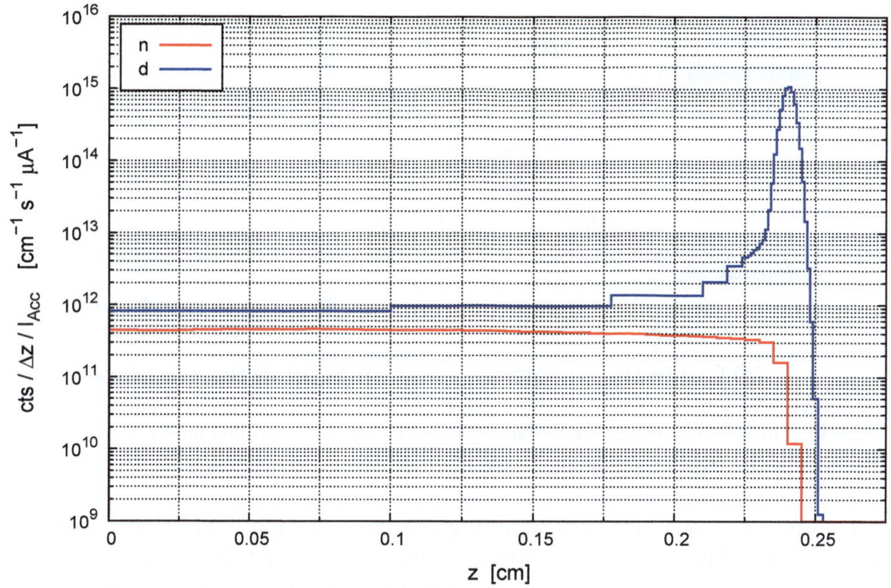

(a) Absolute number of events per unit length.

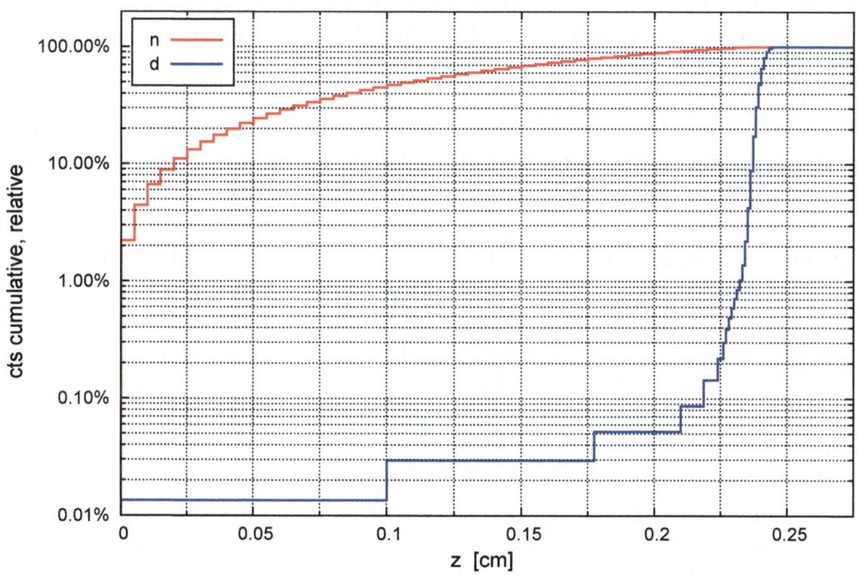

(b) Cumulative relative representation.

Fig. 4.14 Neutron production and deuteron termination depending on the depth inside a beryllium bulk target for 25 MeV deuterons

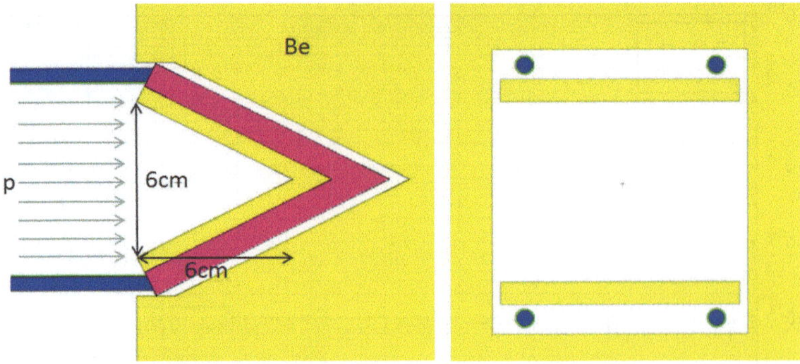

(a) Cut in the z-x-plane. **(b)** Cut in the x-y-plane.

Fig. 4.15 Simulation model of the wedge target embedded by a beryllium moderator

Due to the vague situation regarding the cross section data of deuterons, the following simulations concern a 50 MeV proton accelerator. The accelerator power is chosen to be 100 kW providing a good neutron yield and keeping reasonable target dimensions. The active surface of the target amounts to 100 cm^2 resulting in a surface power density of 1 kW/cm^2 which is considered as the upper limit for water-cooled devices [18–22]. Nonetheless the target should be as compact as possible in order to keep the maximum value of the thermal neutron flux in the moderator as high as possible (cf. Sect. 3.4). For this reason the active target material is not shaped as a plane layer. Instead, the target layer is divided into two parts which are arranged like a wedge as depicted in Fig. 4.15. This idea is adopted from studies on the development of an accelerator-based BNCT facility at Legnaro, Italy [23, 24].

The smallest cube which may encompass the irradiated surface of the wedge has an edge length of 6.7 cm. This results in the opening angle of the wedge to be $\alpha = 53.2°$. As the angle between proton beam and wedge surface is given by $\alpha/2 = 26.6°$ and the optimum proton path length in beryllium amounts to 1.4 cm according to Sect. 4.1.2.2, Fig. 4.9a. Thus, the beryllium layer thickness is calculated to be $d_{Be} = 1.4\,\text{cm} \cdot \sin \alpha/2 = 0.626\,\text{cm}$. The thickness of the backing plate is chosen to be 1 cm. Accordingly, the outer target dimensions are $d_x = 8.9\,\text{cm}$, $d_y = 7.5\,\text{cm}$, and $d_z = 9.6\,\text{cm}$. The inlet and outlet pipes ($r_{outer} = 0.3\,\text{cm}$ and $r_{inner} = 0.25\,\text{cm}$) of the cooling circuit are attached on the outward corners of the wedge (negative z-direction) to the backing plate, as can be seen in Fig. 4.15. Explicit modeling of the cooling tubes in the backing plate is omitted for simplicity. Instead, the coolant is assumed to be distributed homogeneously in the material of the backing plate with a volume fraction of 20%. This approach has no significant influence on the neutronic performance of the target-moderator-reflector assembly.

Also a planar target with a surface area of $10 \times 10\,\text{cm}^2$ is examined. The thickness of the beryllium layer amounts to 1.4 cm and of the backing plate to 1 cm. The inlet and outlet pipes of the cooling circuit are omitted and the coolant is homogeneously distributed in the backing plate with a volume fraction of 20%. For comparative

Table 4.3 Maximum thermal flux, its relative error, and the gain in thermal flux with respect to the wedge inside a light water moderator with wedge, planar, and reverse wedge target

Target shape	Φ_{th}^{max} $(s^{-1}\,cm^{-2}\,\mu A^{-1})$	$\sigma_{rel}(\Phi_{th}^{max})$ (%)	Gain
Wedge	8.35×10^8	0.9	–
Planar	9.05×10^8	1.2	+8%
Reverse wedge	8.79×10^8	0.9	+5%

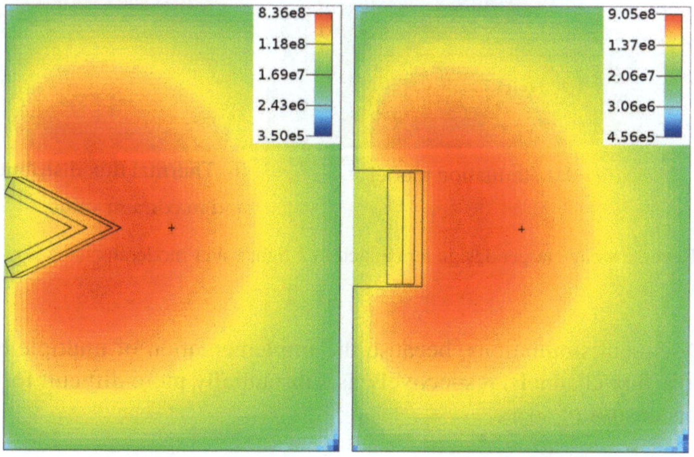

(a) Wedge target. **(b)** Planar target.

Fig. 4.16 Thermal flux distribution and maximum values in light water-based moderators with different target shapes

simulations the planar target and the wedge target are investigated by means of the maximum value of the thermal neutron flux when embedded inside a cylindrical light water moderator with radius $r = 20$ cm and height $h = 30$ cm. The flux distribution is recorded by a mesh tally over the whole geometry with a resolution of 0.5 cm. Results and geometry are depicted in Fig. 4.16 and Table 4.3, respectively.

Surprisingly, the planar target delivers a higher maximum value of the thermal flux. This can be understood by considering the shape of the thermal neutron cloud (dark red region in Fig. 4.16) which looks like a kidney clung to the wedge target while the cloud for the planar target is more compact, and thus leads to a higher maximum flux value.

For this reason a further target design is taken into consideration which is basically a reverse wedge geometry. The idea is to prevent the target wedge tip from penetrating the thermal neutron cloud but instead to envelop the cloud with the target halves. The geometry is depicted in Fig. 4.17a and the corresponding thermal neutron flux distribution when embedded inside the aforementioned light water moderator is given in Fig. 4.17b. Although the planar target and the reverse wedge target deliver slightly higher thermal neutron fluxes in the moderator, the original wedge design is used

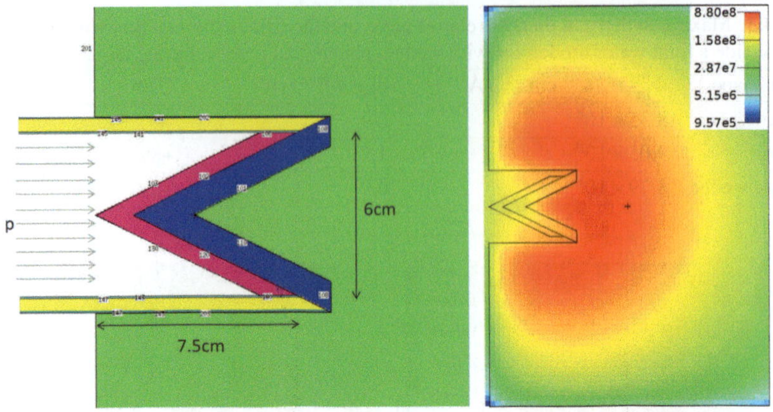

(a) Simulation model. (b) Thermal flux distribution and maximum value.

Fig. 4.17 Reverse wedge target design embedded in a light water moderator

for all subsequent simulations, because the implementation of multiple beamlines and multiple flux channels, respectively, is substantially more difficult for the other designs for reasons of space.

4.1.4 Materials

The same approach is followed as at RANS: the irradiated beryllium is placed on a backing plate, which absorbs the deposited ions. Simulation studies are performed to investigate the impact of choosing different coolants as well as different backing plate materials. Results are explained hereafter.

Influence of Different Cooling Media Water is a typical coolant. Additionally, also gallium or eutectic gallium alloys can be used, which can remove an amount of heat approximately three times as large, due to its higher thermal conductivity [18, 25], see Table 4.4 . The structure of a gallium cooling device is more complex, though commercially available. Liquid gallium is compatible with stainless steels, titanium, beryllium, and water but is corrosive against some metals, e.g., aluminum [18]. Besides water and pure gallium, an eutectic alloy of 75% gallium and 25% indium and an alloy called Galinstan (68.5% Ga + 21.5% In + 10% Sn) [26] are also investigated. The coolant is homogeneously distributed in the backing plate with a volume fraction of 20%. Comparison is done by means of the maximum and average thermal flux inside a big ($r = 100\,\text{cm}$, $l = 200\,\text{cm}$) and a small ($r = 20\,\text{cm}$, $l = 30\,\text{cm}$) H_2O moderator.

According to the results given in Fig. 4.18 there are virtually no differences between water and pure gallium. For the two eutectic alloys the maximum thermal flux is about 5% lower. The average flux throughout the whole moderator volume is

Table 4.4 Density, melting and boiling temperature, and thermal conductivity of cooling media considered for simulation

Parameter	H_2O	Ga	75% Ga, 25% In	Galinstan
ρ [$g\,cm^{-3}$]	1	5.9 [27]	6 [28]	6.44 [29]
T_{melt} [°C]	0	29.8 [27]	15.5 [28, 30]	−19.5 [26]
T_{boil} [°C]	100	2400 [31]		>1300 [29]
$\lambda_{thermal}$ [$W\,m^{-1}\,K^{-1}$]	0.6	29 [27]		16.5 [29]

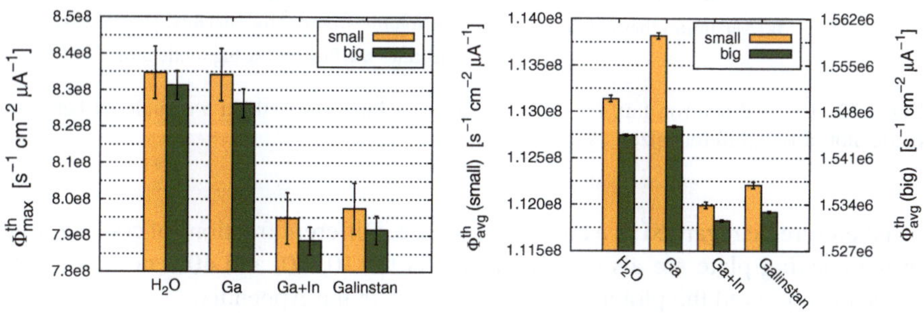

(a) Maximum thermal flux.

(b) Volume-averaged thermal flux. The right axis label indicates the flux for the big moderator.

Fig. 4.18 Thermal neutron flux for different coolants in the backing plate of the wedge target for a cylindrical beryllium moderator of two different sizes

less than 1% lower which is explained by the relatively high absorption cross section of indium. Therefore, the absorption cross section of the alloys is an order of magnitude larger in the thermal region than that of gallium as presented in Fig. A.3 in the Appendix.

Target Backing Materials Besides of aforementioned vanadium, also niobium, titanium, palladium, and platinum are suitable materials for the backing plate as discussed in Sect. 2.2.2. These materials are investigated by means of the maximum and average thermal flux inside a big ($r = 100\,cm$, $l = 200\,cm$) and a small ($r = 20\,cm$, $l = 30\,cm$) H_2O moderator. The cooling water is homogeneously distributed in the backing plate with a volume fraction of 20%.

The influence of the backing plate material is relatively low, as indicated by Fig. 4.19. Regarding neutronic performance platinum is preferable. Apart from the volume-averaged flux for small moderators the advantage over niobium and zirconium is negligible. When using vanadium flux values are about 5–10% lower which can be explained by the the shape of the cross section given in Fig. A.4 in the Appendix. The resonance region of vanadium extends up to 5 MeV and has very broad peaks at energies around 10 keV. Therefore, a lot of fast and epithermal neutrons are absorbed in the backing plate, because in the vicinity of the target the flux of fast and epithermal neutrons is relatively high. Since it has been used for the target of RANS, vanadium is also used in all further simulations as backing plate material and

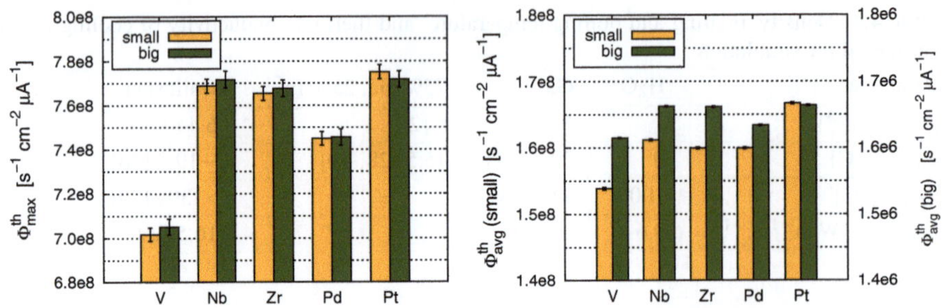

(a) Maximum thermal flux.

(b) Volume-averaged thermal flux. The right axis label indicates the flux for the big moderator.

Fig. 4.19 Thermal neutron flux for different target backing materials in a cylindrical beryllium moderator of two different sizes

can therefore be regarded as a conservative case. Different fractions of cooling water in the backing plate are also investigated. As for 10, 30, and 40% no noteworthy impact is observed the plot are given in Fig. B.5 in the Appendix.

4.2 Moderator Development

The influence of the target design is always studied with compact and quasi-infinite light water moderators. In order to optimize the moderator itself, the wedge target design described in Sect. 4.1.3 is used. Vanadium with cooling water with a volume fraction of 20% is employed as the backing plate material. First, typical moderator materials are investigated. Subsequently, the outer shape of the moderator is optimized. Based on these two studies, a moderator will be developed, which is a combination of different materials. Subsequently flux fingers are implemented in the moderator system, which are partly provided in a further step with cold moderators based on parahydrogen and methane.

4.2.1 Choice of Moderator Materials

The mechanism of moderation is already outlined in Sect. 2.3. Typical moderator materials therefore exhibit a very high elastic scattering cross section, the largest possible energy transfer per collision and thus the lowest possible nuclear mass, and a very low absorption cross section.

Apart from the theoretical considerations, the following materials were tested for suitability as a moderator with MCNP: light water, heavy water, beryllium, graphite and polyethylene. Compact and quasi-infinite cylindrical moderators com-

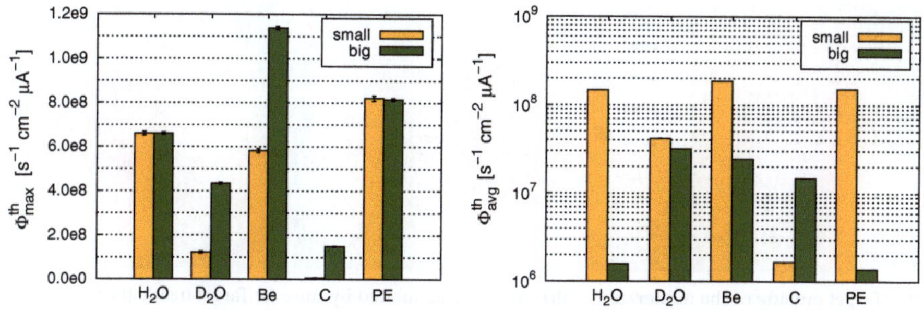

(a) Maximum thermal flux.

(b) Volume-averaged thermal flux. This is a logarithmic plot.

Fig. 4.20 Thermal neutron flux for different moderator materials with cylindrical shape of two different sizes

prising these materials are investigated concerning the maximum thermal flux and the volume-aver-aged flux. The results are depicted in Fig. 4.20. For the hydrogen-based moderators based on H_2O and PE the maximum flux of thermal neutrons is identical for both volumes and is already saturated even in small volumes due to the high slowing-down power. The volume-averaged flux for small volumes is larger by two orders of magnitude than for large volumes due to the relatively high absorption cross section of hydrogen.

Heavy water offers a lower slowing-down power, which is why the maximum flux is significantly lower for the small configuration. The average flux is nearly identical for both volumes because of the extremely small absorption cross section.

The behavior of the beryllium moderators is comparable to the moderators of heavy water. However, due to the high (n, xn) cross section higher thermal fluxes are achieved. The absorption cross section is slightly higher than that of heavy water, which is why the volume-averaged flux is almost one order of magnitude higher in the small moderator than in the large one.

Graphite offers a very low maximum thermal flux even for large volumes because of its low slowing-down power. However, due to the extremely small absorption cross section, cf. Fig. A.1 in the Appendix, a high average flux can be achieved at least with a large volume.

4.2.2 Target Position

It can be assumed that a significant amount of fast neutrons is backscattered in the first layer of the moderator after leaving the target and is therefore lost. This leakage can be reduced by enclosing the target by the moderator material as completely as possible. In a parametric study the maximum value of the thermal flux is determined for different positions in the moderator. Figure 4.21 show the corresponding simulation

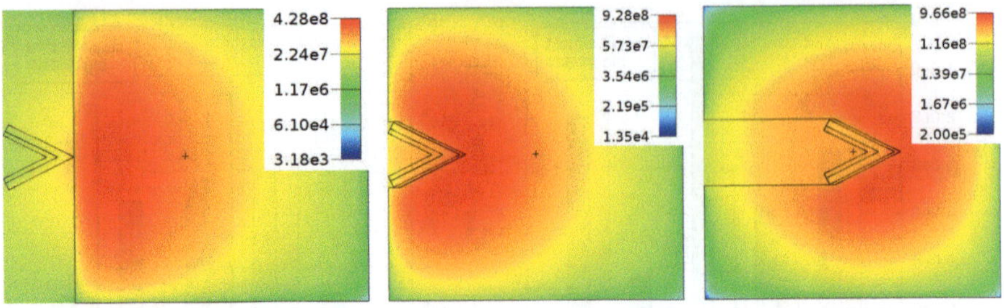

(a) Target outside of the moderator. **(b)** Target encapsulated by the moderator. **(c)** Target inside the moderator.

Fig. 4.21 Simulation model and flux distribution of thermal neutrons of a cylindrical beryllium moderator with a volume of 50 dm^3 and different positions of the wedge target. Unit of the flux is $cm^{-2}\,s^{-1}\,\mu A^{-1}$

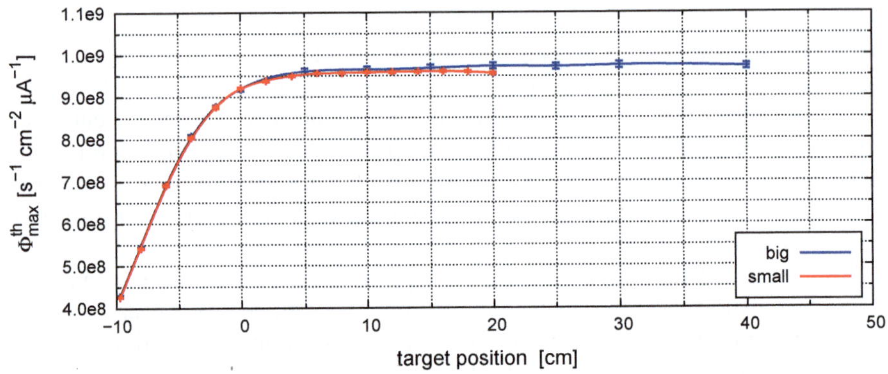

Fig. 4.22 Maximum thermal flux of a small and a quasi-infinite beryllium moderator depending on the target position. −9.71 cm corresponds to the position depicted in Fig. 4.21a and 0 cm corresponds to Fig. 4.21b

model and the calculated flux distribution of thermal neutrons. This is again done for a compact and a quasi-infinite beryllium moderator. The results are summarized in diagram 4.22.

Self-evidently, the leakage of neutrons is the larger, the farther the target is located from the moderator. Consequently, the configuration presented in Fig. 4.21a exhibits the lowest maximum thermal flux. Compared to this, the maximum value of the thermal flux is twice as high if the target is located in the moderator so far that both flanks are just now enclosed by moderator material as depicted in Fig. 4.21b. If the target is further moved in, the maximum value of the thermal flux no longer increases significantly. The average flux however continues to increase. The reason is that the area of high thermal fluxes is shifted to the center of the moderator for deeper target positions, which is evident from Fig. 4.21. Therefore, the thermal flux in peripheral areas is also significantly larger leading to a much higher average flux for the whole volume. In contrast the maximum value for target positions larger than 6 cm remains almost constant.

Since configurations in which the target is completely enclosed by the moderator are very difficult to realize from a mechanical engineering point of view, the configuration with target position 0 cm is used in all further studies if not stated otherwise.

4.2.3 *Shape, Dimensions and Geometry*

In Sect. 3.3.1 the effect of choosing different shapes (sphere, cylinder, cap, ellipsoid, and cone) for a compact moderator of a laser-driven device are investigated. The highest thermal neutron flux is achieved in a cap configuration, which consists of a cylinder and a superimposed half-sphere, and cylindrical moderators. Hence these two shapes are taken into consideration for the optimization study regarding the moderator dimensions for a 50 MeV CANS.

Parameter studies are performed in which the maximum value of the thermal flux distribution is determined in dependence on the moderator volume. The investigations are carried out for polyethylene, beryllium, and graphite. Since the volume depends on two parameters, namely the radius r and the height h, both parameters are varied independently. The results are shown in Fig. 4.23. For each point in the diagram only the flux value of the optimum combination of r and h is shown. Since the materials

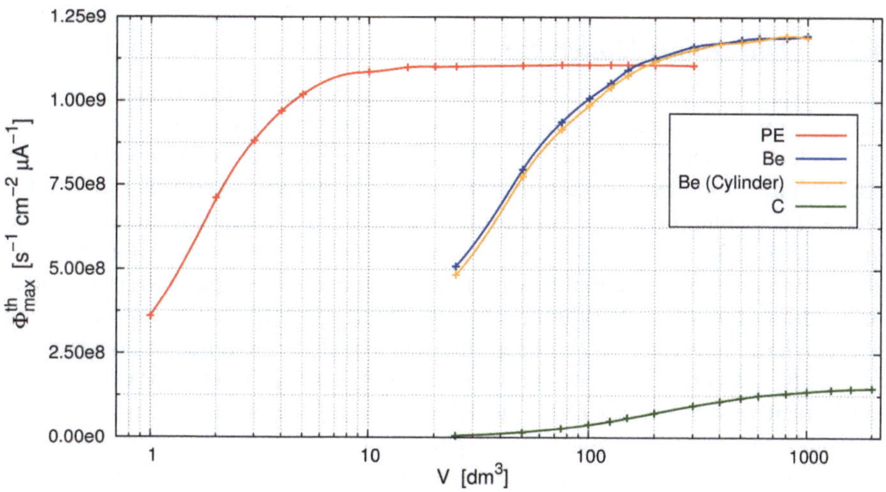

Fig. 4.23 Maximum value of the thermal flux for different moderator materials depending on their volume. Because the volume of the underlying geometries depend on two parameters, radius r and height h, only the optimum combination of these are chosen for each datapoint. The results for cylindrical shapes of polyethylene and graphite are almost identical to the "cap"-configuration

investigated in Sect. 4.2.1 exhibit different moderation times and lengths, cf. Sect. 2.3 there is a quite large dependence on the dimensions of the moderator. For this reason, the optimum geometries for polyethylene, beryllium, and graphite are determined.

For each of the materials the maximum thermal flux becomes saturated above a certain moderator volume. For polyethylene saturation is reached with a flux value of $1.1 \times 10^9\,\mathrm{cm^{-2}\,s^{-1}\,\mu A^{-1}}$ already at a volume of about $15\,\mathrm{dm^3}$ due to the extremely large slowing-down power. With beryllium the maximum flux amounts to $1.2 \times 10^9\,\mathrm{cm^{-2}\,s^{-1}\,\mu A^{-1}}$ which is achieved above a volume of about $500\,\mathrm{dm^3}$ (the maximum value of polyethylene is met at a beryllium volume of $200\,\mathrm{dm^3}$), indicating that beryllium has a better moderating ratio despite the much lower slowing-down power. On the one hand, this is due to the much lower absorption cross section and, on the other hand, due to the (n, 2n) reaction occurring in beryllium. With graphite very large volumes larger than $1\,\mathrm{m^3}$ are required to achieve a relatively small maximum value of only $1.5 \times 10^8\,\mathrm{cm^{-2}\,s^{-1}\,\mu A^{-1}}$. Although graphite has a very low absorption cross section, the moderating ratio is significantly lower than for polyethylene and beryllium because of the low slowing-down power.

Above the saturation volume as many neutrons are absorbed, as are backscattered to the center of the moderator. It is therefore reasonable to use a small volume moderator of polyethylene surrounded by a reflector made of beryllium. In this case, thermal neutrons, leaving the polyethylene, are backscattered from the beryllium without being absorbed (which would be the case for a larger polyethylene moderator). In the next section studies are presented in which various combinations of moderator and reflector materials for different dimensions are treated.

4.2.4 The Onion Moderator

In the first step, a simple combination of polyethylene as a moderator and beryllium as a reflector is examined, see Fig. 4.24. The dimensions are chosen such that the total volume of moderator and reflector amounts to $200\,\mathrm{cm^3}$. According to Fig. 4.23 this is the volume of a pure beryllium moderator which delivers the same maximum thermal flux, as would be possible with a polyethylene moderator. With this choice, the maximum thermal flux in the onion moderator is therefore independent of the dimension of the beryllium reflector. Thus, it is sufficient to vary the size of polyethylene moderator and to keep the dimensions of the reflector constant. For simplicity, a spherical shape is chosen for the polyethylene moderator whose center is identical with the vertex of the beryllium layer of the target, cf. Fig. 4.24a. The minimum radius of the polyethylene sphere is limited to $r_{\mathrm{PE}} = 8.5\,\mathrm{cm}$ due to the dimensions of the wedge target. Hence, smaller PE volumes can be examined only if the polyethylene is modeled as a thin layer behind the target area as shown in Fig. 4.24b. In Fig. 4.25 the maximum thermal flux is plotted as a function of the volume of the polyethylene. Volumes up to $1.6\,\mathrm{cm^3}$ correspond to a polyethylene layer according to Fig. 4.24b; larger volumes correspond to a spherical shape (Fig. 4.24a). For comparison, the curves for pure PE and Be moderators are shown, taken from

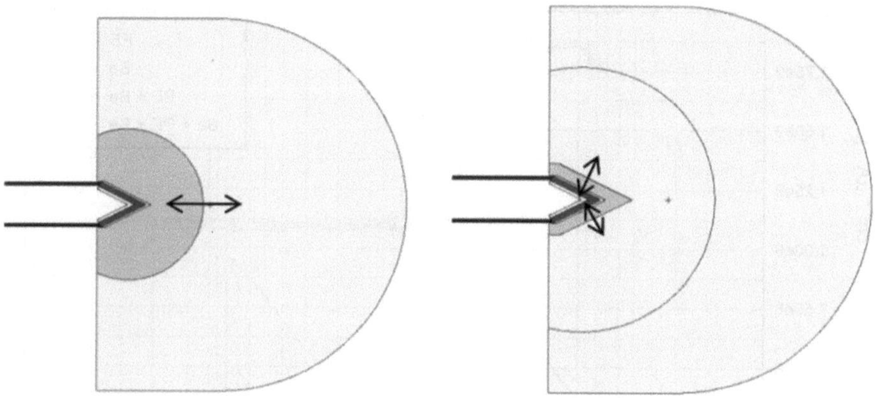

(a) PE moderator with spherical shape **(b)** Wedge-shaped PE moderator to realize smaller volumes

Fig. 4.24 Simulation model for estimating the thermal flux distribution of a moderator-reflector assembly with the wedge target for different volumes of the PE moderator. Materials from left to right: water (inside the tubes), beryllium, vanadium, polyethylene, and beryllium

Fig. 4.23. Obviously, the maximum thermal flux can be significantly increased by combining a small-volume polyethylene moderator with a large beryllium reflector. For thicknesses $\ll 1\,\text{cm}$ the same maximum flux as with a pure Be moderator is achieved. With increasing layer thickness the maximum thermal flux increases up to a value of $1.87 \times 10^9\,\text{cm}^{-2}\,\text{s}^{-1}\,\mu\text{A}^{-1}$ at a PE volume of $1\,\text{dm}^3$ which corresponds to a layer thickness of 2.2 cm. For larger volumes the maximum flux decreases again until the same value as that of a pure PE moderator is achieved ($1.1 \times 10^9\,\text{cm}^{-2}\,\text{s}^{-1}\,\mu\text{A}^{-1}$) at a PE volume of about $10\,\text{dm}^3$. By combining PE and beryllium, the maximum thermal flux can therefore be increased by 68% and 56% compared to a pure PE or Be moderator, respectively. Furthermore, the effect of an additional layer of beryllium between target and PE moderator is investigated. At first, the layer thickness of the first beryllium layer is varied. Afterwards a second study is conducted in which the thickness of the PE sphere is varied and optimized. The results for this study is represented by the orange line in Fig. 4.25. It is evident that the maximum flux for those configurations is less than that of a PE-Be assembly which is explained by the fact that although beryllium exhibits a high (n, 2n) cross section and the flux density of fast neutrons directly behind the target is maximal, the compactness of the PE moderator is disturbed by the additional beryllium layer such that the maximum thermal flux is lower.

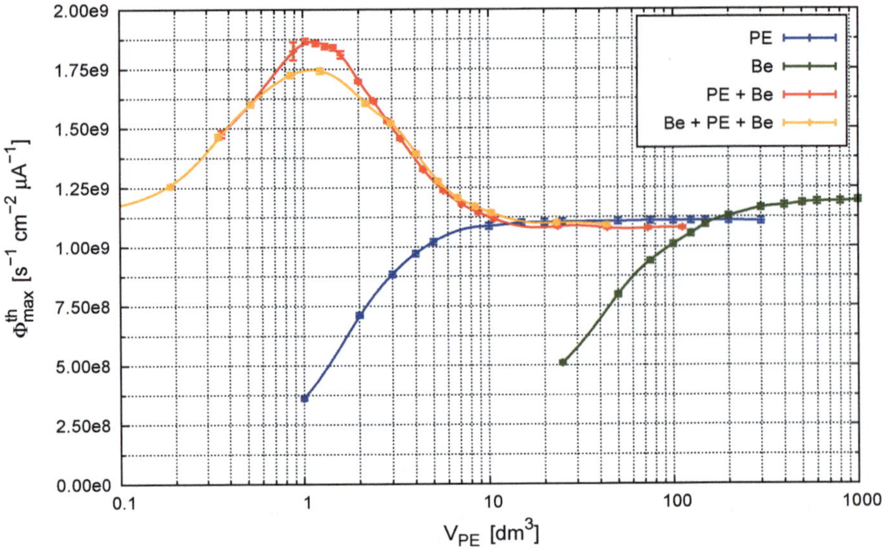

Fig. 4.25 Value of the maximum thermal flux inside the moderator (-reflector) assembly for a pure PE moderator, a pure Be moderator, and a PE moderator with beryllium reflector in dependence to the volume of the PE moderator or the Be moderator, respectively

4.2.5 Flux Channels and The Finger Moderator

Up to now, the moderator has been optimized to maximize the maximum thermal flux at its center. As already discussed in Sect. 3.4, the thermal neutrons are to be extracted from the flux maximum by means of flux channels. In the flux maximum, not only is the thermal total flux significantly larger, but also the radial component of the particle current $\vec{j}_{\vec{r}}$ of thermal neutrons. In addition, the effective surface area from which the neutrons are extracted is increased by the area of the inner surface of the flux channels.

Since the HBS is mainly used to investigate samples of sizes $1-10\,\text{cm}^2$, five flux channels of radius $r = 1$ cm and two channels of radius $r = 3$ cm are implemented into the moderator-reflector assembly. The positions are selected in such a way that the origins of the channels equally distributed in the center of the moderator. This results in the angle of incidence relative to the rotation axis of the moderator of $\varphi = 10°, 45°, 90°, 120°$ for $r = 1$ cm and $\varphi = -10°, -90°$ for $r = 3$ cm. To investigate the performance of producing cold neutrons, the channel with $r = 1$ cm and $\varphi = 45°$ is equipped with a cold moderator consisting of parahydrogen at a temperature of $T = 20$ K and the channel with $r = 3$ cm and $\varphi = 90°$ with solid methane at $T = 22$ K. For these radii the lengths of the cold moderators are $l = 4$ cm (para) and $l = 2.5$ cm (methane) based on the simulation results from Sect. 3.3.3, Table B.1. On the basis of the final design shown in Fig. 4.26, the brilliance, the temporal pulse structure as well as the spectrum are examined. In addition, the dependencies of these

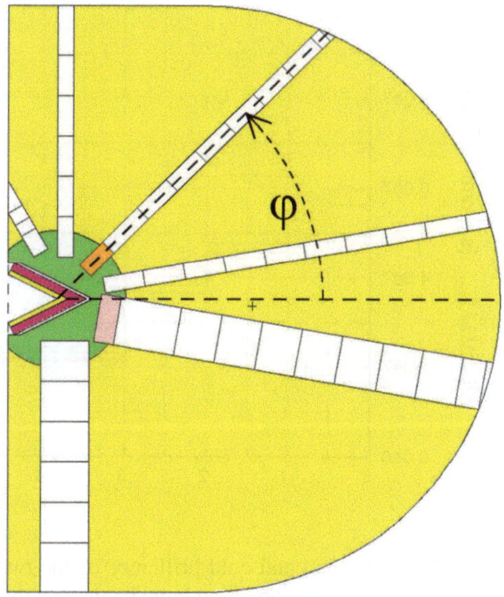

Fig. 4.26 Simulation model of the final design with wedge target, polyethylene moderator, beryllium reflector, small and large flux channels, a cold parahydrogen moderator (orange), and a solid methane moderator (salmon)

quantities on the thickness of the reflector layer as well as on the pulse structure of the accelerator are investigated [32].

Multiple Flux Channels Similar to to Sect. 3.4.4 for a laser-driven system, investigations are conducted to answer the question if multiple flux channels interfere with each other. It would be conceivable that the peak brilliance at the channel exits is significantly lower because, on the one hand, moderator material is lost, which is mainly due to the loss of PE and, on the other hand, the compactness of the PE moderator is disturbed. The fact that the latter case has a negative effect on the peak brilliance has already been dealt with in Sect. 4.2.4.

The behavior of a system with only two flux channels (reduced model) with $r = 1\,\text{cm}$, $\varphi = 90°$ and $r = 3\,\text{cm}$, $\varphi = 10°$, whereby the latter contains a cold methane moderator, is compared to the brilliance of the same channels in the full model. The results are shown in Fig. 4.27 and Table 4.5. It can be seen that the peak brilliance in the full model is lower by 5.8% for the thermal channel and by 3.9% for the cold channel.

It can be summarized that multiple channels are only slightly interfering with one another by reducing the moderator volume.

Comparison with a Small Target As a supplement to the investigations of different target geometries in Sect. 4.1.3, a nearly point-like target is simulated to determine how the size of the target affects the maximum thermal flux in the moderator and the peak brilliance at the channel exits. Similar to Sect. 4.2.4, the maximum thermal flux for a pure PE moderator is determined as a function of its volume, as depicted in Fig. 4.28a, and subsequently for a PE moderator with a beryllium reflector, for which the outer dimensions are adjusted to give the moderator assembly a total volume of $200\,\text{dm}^2$, see Fig. 4.28b.

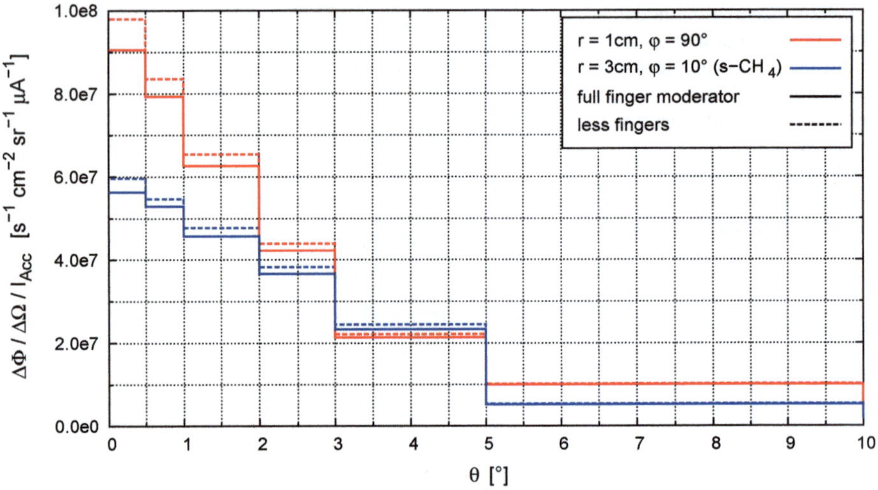

Fig. 4.27 Thermal and cold brilliance of the two channels in the reduced model (dashed lines) compared to the the corresponding channels in the full model (solid lines)

Table 4.5 Thermal and cold peak brilliance for $\theta \leq 1°$ and relative error at the flux channel exit for full-scale model and a model with only two channels. Unit of the differential flux is $s^{-1}\,cm^{-2}\,sr^{-1}\,\mu A^{-1}$. The loss in peak brilliance is given with respect to the reduced model

Finger position	$\frac{\Delta\Phi^{full}}{\Delta\Omega\, I_{Acc}}$	σ_{rel} (%)	$\frac{\Delta\Phi^{reduced}}{\Delta\Omega\, I_{Acc}}$	σ_{rel} (%)	Loss (%)
$r = 1\,cm, \varphi = 90°$	8.34×10^7	3.1	8.85×10^7	2.0	−5.8
$r = 3\,cm, \varphi = 10°$	5.46×10^7	3.1	5.68×10^7	0.8	−3.9

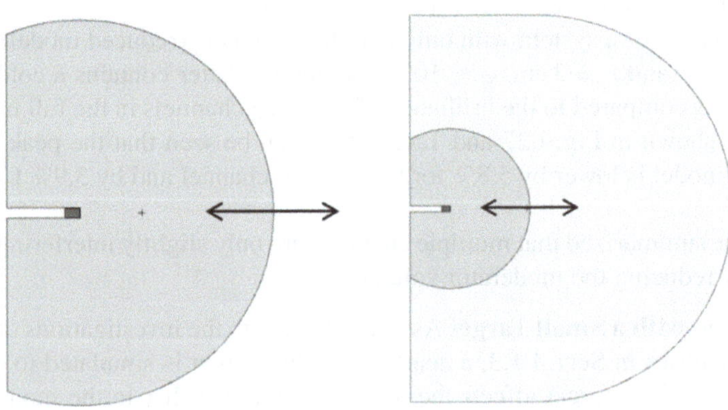

(a) Pure PE moderator

(b) PE moderator with beryllium reflector

Fig. 4.28 Simulation model for estimating the thermal flux distribution of a moderator (-reflector) assembly with a point-like target for different volumes of the PE moderator

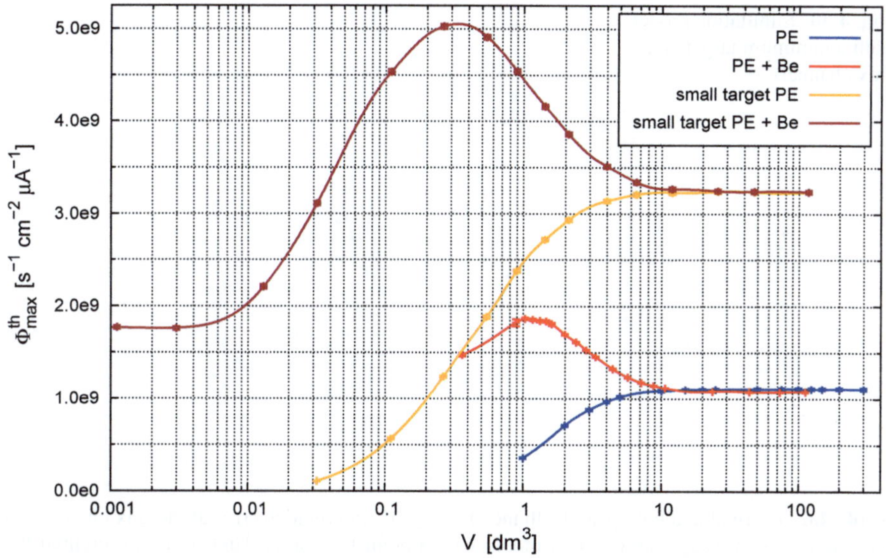

Fig. 4.29 Value of the maximum thermal flux inside the moderator (-reflector) assembly for a pure PE moderator, a PE moderator with beryllium reflector and both configurations with a small target in dependence to the volume of the PE moderator

Results are shown in Fig. 4.29 together with the maximum thermal flux for a moderator assembly with a wedge target. With a pure PE moderator the saturation of the maximum thermal is reached at $V = 10\,\text{dm}^3$ of PE with a value of $\Phi_{\text{th}}^{\text{max}} = 3.25 \times 10^9\,\text{cm}^{-2}\,\text{s}^{-1}\,\mu\text{A}^{-1}$, which is a factor of 2.95 larger than for the assembly with wedge target ($\Phi_{\text{th}}^{\text{max}} = 1.1 \times 10^9\,\text{cm}^{-2}\,\text{s}^{-1}\,\mu\text{A}^{-1}$).

For the polyethylene moderator with beryllium reflector the maximum value of the maximum thermal flux is reached at a PE volume of $V = 0.3\,\text{dm}^3$ with $\Phi_{\text{th}}^{\text{max}} = 5.0 \times 10^9\,\text{cm}^{-2}\,\text{s}^{-1}\,\mu\text{A}^{-1}$ which is a factor of 2.66 larger than for the assembly with wedge target ($\Phi_{\text{th}}^{\text{max}} = 1.9 \times 10^9\,\text{cm}^{-2}\,\text{s}^{-1}\,\mu\text{A}^{-1}$). Interestingly, the volume needed to reach the maximum value is much lower compared to the wedge assembly. This is due to the fact, that the PE moderator is much more compact with a point-like target so that the optimum PE volume is lower. On the basis of the results shown, it is to be expected that not only the thermal flux in the moderator but also the peak brilliance at the channel exits is significantly higher with point-shaped targets.

In simulations using the geometry depicted in Fig. 4.30 the brilliance at the channel exits are determined. The results are shown in Fig. 4.31 which also encompasses the brilliance of the configuration with a wedge target according to Fig. 4.26. The values of the peak brilliance in forward-direction are additionally summarized in Table 4.6. The largest deviation of 64% occurs at the channel $r = 1$ cm, $\varphi = 120°$ whereas the channel $r = 3$ cm, $\varphi = 10°$ exhibits the lowest deviation of 9%.

The deviation of the peak brilliance between the point-like target and the wedge target is less pronounced than their deviation in the maximum thermal flux since the

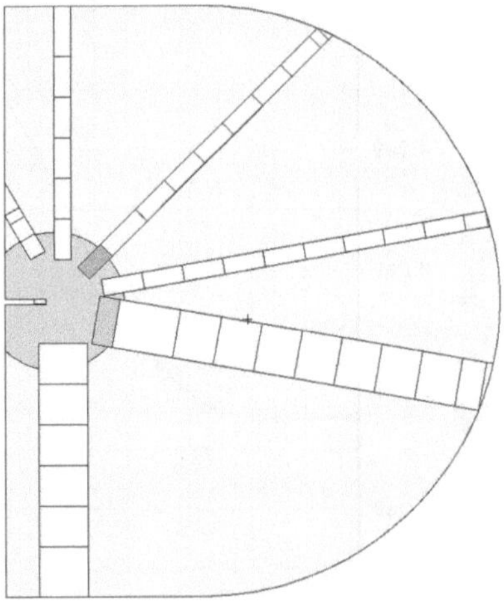

Fig. 4.30 Simulation model with a minimum target and flux channels

Table 4.6 Thermal and cold peak brilliance for $\theta \leq 1°$ and relative error at the flux channel exit for a model with wedge target and a model with a point-like target. Unit of the differential flux is $\mathrm{cm}^{-2}\,\mathrm{s}^{-1}\,\mathrm{sr}^{-1}\,\mu\mathrm{A}^{-1}$. The gain in peak brilliance by usage of a point-like target is given with respect to the full model

Channel position	$\frac{\Delta\Phi^{\mathrm{wedge}}}{\Delta\Omega\, I_{\mathrm{Acc}}}$	$\sigma_{\mathrm{rel}}\,(\%)$	$\frac{\Delta\Phi^{\mathrm{point}}}{\Delta\Omega\, I_{\mathrm{Acc}}}$	$\sigma_{\mathrm{rel}}\,(\%)$	Gain (%)
$r = 1\,\mathrm{cm}, \varphi = 10°$	8.31×10^7	1.9	1.02×10^8	2.6	23.1
$r = 1\,\mathrm{cm}, \varphi = 45°$	8.32×10^7	1.8	9.34×10^7	2.6	12.3
$r = 1\,\mathrm{cm}, \varphi = 90°$	8.62×10^7	1.9	1.15×10^8	2.4	33.3
$r = 1\,\mathrm{cm}, \varphi = 120°$	5.90×10^7	2.3	9.67×10^7	2.7	63.7
$r = 3\,\mathrm{cm}, \varphi = 10°$	5.36×10^7	0.8	5.82×10^7	1.1	8.5
$r = 3\,\mathrm{cm}, \varphi = 90°$	6.01×10^7	0.8	8.43×10^7	1.0	40.1

thermal flux maximum in the moderator is higher but spatially more concentrated for the point-like target. Additionally, the gradient of the thermal flux is higher around the maximum which is why the thermal flux at the bottom of the channels is only slightly higher.

A configuration where the flux channel's origins are located on the surface of the polyethylene sphere and not at the thermal flux maximum inside the PE moderator is investigate. For this setup the peak brilliance at the channel exits is reduced by 10% to 40% depending on the channel as indicated by Fig. B.6. This approach is therefore not pursued in further studies.

4.2.5.1 Spectra

Based on the simulation model in Fig. 4.26, the most important performance parameters of the Finger Moderator are determined. These include the spectra of the neutrons

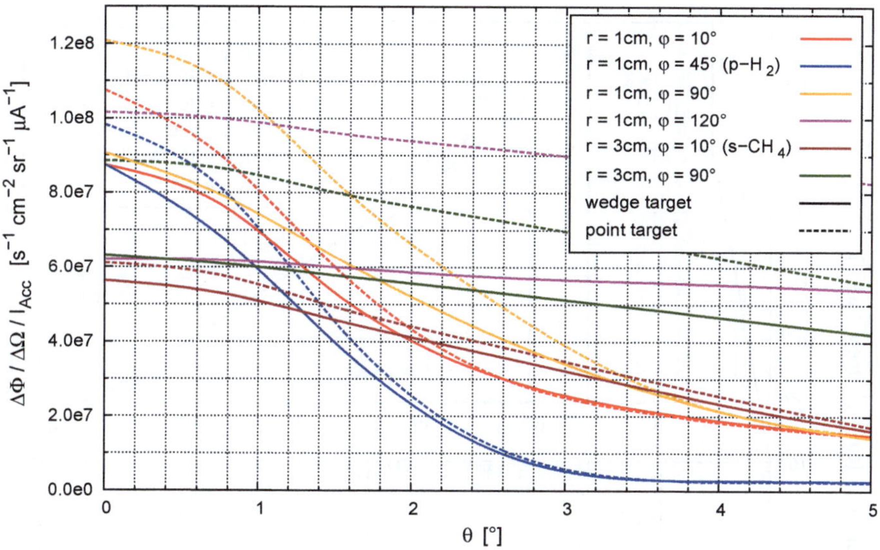

Fig. 4.31 Thermal or cold brilliance, respectively, at the channel exits of the full model with wedge target (solid lines) and for a point-like target (dashed lines)

being extracted, as well as the brilliance and the pulse structure at the channel exits. The spectra of the neutrons passing through the exits of the channels are shown in Fig. 4.32. The upper plot extends over the full energy range, whereas the lower plot highlights the cold and thermal energy range indicated by dashed lines in both diagrams. The absolute flux values, which are integrated over the specified energy ranges, are summarized in Table 4.7.

When looking at the spectra at the cold channel exits, it is noticeable that the peaks are positioned at different energies. The spectrum of the channel with parahydrogen reaches its maximum at an energy of 5.2 meV, which corresponds to a temperature of 60 K. Since the parahydrogen is maintained at a temperature of 20 K, the generated neutrons are considered undermoderated. The spectrum at the cold channel with solid methane exhibits a smooth maximum at an energy of 2.2 meV corresponding to a temperature of 25 K. The neutrons are therefore in thermal equilibrium with the methane, whose temperature amounts to 22 K. Since the decrease of the spectrum of parahydrogen towards lower energies is relatively flat and the maximum appears at higher energies, the total flux at the corresponding channel exit is higher than that of methane ice ($7 \times 10^4\,\mathrm{cm}^{-2}\,\mathrm{s}^{-1}\,\mu\mathrm{A}^{-1}$ vs. $5.13 \times 10^4\,\mathrm{cm}^{-2}\,\mathrm{s}^{-1}\,\mu\mathrm{A}^{-1}$). In addition, the parahydrogen moderator still has a fairly large background of thermal neutrons making it suitable for experiments working with bi-spectral neutron beams, e.g., chopper spectrometers [33, 34].

The shape of the spectra of the thermal flux channels are almost identical. For the two channels with $r = 1\,\mathrm{cm}$, $\varphi = 10°$ and $r = 1\,\mathrm{cm}$, $\varphi = 90°$, respectively, the total thermal flux approximately amounts to $7.7 \times 10^4\,\mathrm{cm}^{-2}\,\mathrm{s}^{-1}\,\mu\mathrm{A}^{-1}$, whereas

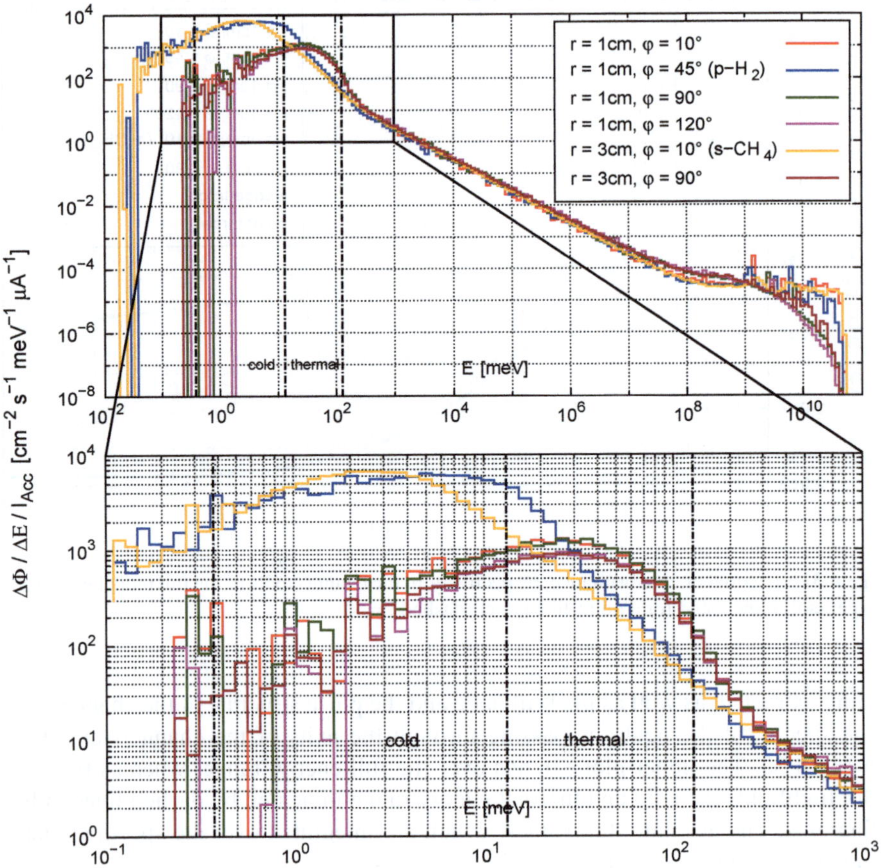

Fig. 4.32 Spectra of neutrons leaving the channels in forward direction ($\theta \leq 1°$)

with the channels $r = 1\,\text{cm}, \varphi = 120°$ and $r = 3\,\text{cm}, \varphi = 90°$ a thermal flux of $5.9 \times 10^4\,\text{cm}^{-2}\,\text{s}^{-1}\,\mu\text{A}^{-1}$ is achieved. It should be noted that the pure yield of thermal neutrons at the exit of the latter channel is significantly higher, since a much larger area has to be integrated over. The low flux at the channel with $r = 1\,\text{cm}, \varphi = 120°$ can be explained by the fact that it can not be placed at the center of the moderator for reasons of space. Consequently, the thermal neutron flux density at the channel bottom remains lower than for the other channels. This is evident by considering Fig. 4.33 which shows the flux distribution for thermal (Fig. 4.33a) and cold (Fig. 4.33b) neutrons. The thermal flux distribution also shows that the comparatively large amount of PE moderator material is removed by the two channels with $r = 3\,\text{cm}$, which significantly reduces the thermal flux below the target. As a result, the thermal flux at $r = 3\,\text{cm}, \varphi = 90°$ is as small as for the channel $r = 1\,\text{cm}, \varphi = 120°$. By comparing the cold and thermal flux distributions, it is easy to recognize the bi-spectral energy

Table 4.7 Flux of cold and thermal neutrons at the exit of the flux channels in forward direction with $\theta \leq 1°$

Channel position	$\Delta\Phi^{\text{cold}}/I_{\text{Acc}}$	$\Delta\Phi^{\text{thermal}}/I_{\text{Acc}}$
	$\text{cm}^{-2}\,\text{s}^{-1}\,\mu\text{A}^{-1}$	$\text{cm}^{-2}\,\text{s}^{-1}\,\mu\text{A}^{-1}$
$r = 1\,\text{cm}, \varphi = 10°$	8.32×10^3	7.55×10^4
$r = 1\,\text{cm}, \varphi = 45°$	7.00×10^4	5.82×10^4
$r = 1\,\text{cm}, \varphi = 90°$	7.89×10^3	7.82×10^4
$r = 1\,\text{cm}, \varphi = 120°$	5.52×10^3	5.88×10^4
$r = 3\,\text{cm}, \varphi = 10°$	5.13×10^4	2.75×10^4
$r = 3\,\text{cm}, \varphi = 90°$	5.52×10^3	5.85×10^4

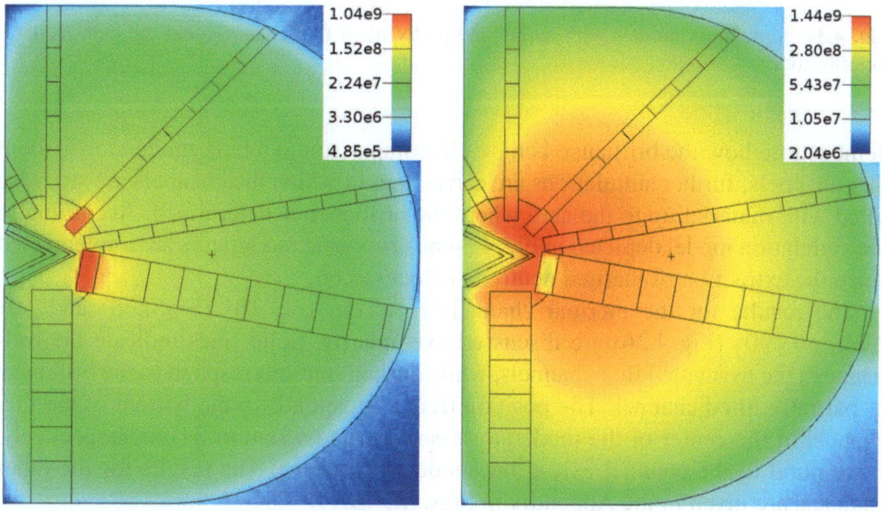

(a) Flux distribution of cold neutrons (0.376 meV – 13.1 meV).

(b) Flux distribution of thermal neutrons (13.1 meV – 127 meV).

Fig. 4.33 Neutron flux distribution measured in $\text{cm}^{-2}\,\text{s}^{-1}\,\mu\text{A}^{-1}$ in the simulation geometry according to Fig. 4.26

distribution of the cold parahydrogen channel, while the methane moderator acts as a sink for thermal neutrons.

It is also striking that the background of fast neutrons with $E > 10^9$ meV is significantly higher for channels with $\varphi > 90°$ than for channels with $\varphi \leq 90°$. However, the latter have a higher background of neutrons with energies 10^7 meV $< E < 10^9$ meV.

4.2.5.2 Brilliance Extraction

In order to show that the thermal brilliance can be extracted in the forward direction by means of flux channels, a simulation is performed in which the brilliance of thermal neutrons is determined at different depths within the flux channels. To additionally

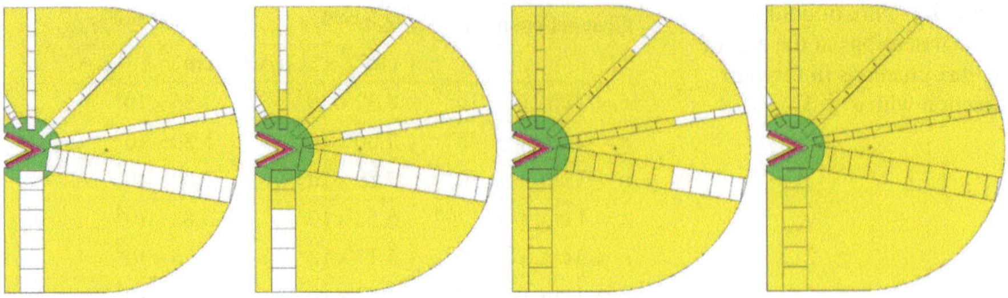

(a) Empty channels (b) Filled 10 cm (c) Filled 30 cm (d) Bulk moderator

Fig. 4.34 Simulation models with empty and partly filled flux channels and the bulk moderator/reflector

demonstrate how the brilliance behaves in comparison to a bulk moderator without flux channels, further simulations are carried out in which the channels are stepwise filled with material up to the area where the brilliance is determined. According to the simulation model depicted in Fig. 4.34 no cryogenic moderators are implemented since the extraction of thermal neutrons is focused on.

The results for the thermal channels $r = 1\,\text{cm}, \varphi = 10°$ (Fig. 4.35) and $r = 3\,\text{cm}, \varphi = 90°$ (Fig. 4.36) are discussed exemplarily. Solid lines indicate the brilliance in the evacuated flux channels, while dashed lines correspond to the brilliance of partially filled channel. The position 0 cm corresponds to the bottom of the flux channel in the center of the moderator assembly and 48 cm or 31 cm, respectively, correspond to the channel exit at the moderator surface. The results for the other channels are given in the Appendix in Figs. B.9–B.11.

Each of the curves decrease from a maximum at 0° to zero at 90° resulting from the fact that 90° corresponds to neutrons with flight direction tangentially to the moderator surface. For positions closer to the center of the moderator the decrease of the brilliance is rather flat whereas the brilliance at the channel exit rapidly decreases in the first few degrees and approaches the minimum with a flat tail. This curve shape can be explained by the fact that tally surfaces closer to the center of the moderator also detect neutrons which are emitted from the channel bottom and have a higher divergence, without being scattered by the inner walls of the flux channel. In the case of tally surfaces further outside the contribution to the brilliance for larger angles originates only from the outermost areas of the inner walls of the channels, where the neutron flux is significantly lower.

By comparing the maximum values at 0° it is evident, that neutrons which leave the moderator in forward direction mainly originate from the center of the moderator. This is not the case for bulk moderators where the brilliance in forward direction is perturbed by the additional material in the channel. The peak brilliance is strongly decreasing for increasing depth inside the flux channel whereas for larger angles the brilliance approaches the values for the case of the Finger Moderator.

At this point it can be concluded that the brilliance in forward direction is extracted from the center of the moderator and is almost completely preserved by the flux

Table 4.8 Peak brilliance of thermal or cold neutrons, respectively, with $\theta \leq 1°$ and its relative error at the flux channel exit and on the bulk moderator surface. Unit of the differential flux $\Delta\Phi$ is $\mathrm{cm^{-2}\,s^{-1}\,sr^{-1}\,\mu A^{-1}}$. The ratio of the peak brilliances of finger and bulk moderator is given as the gain factor

Channel position	$\frac{\Delta\Phi^{\text{finger}}}{\Delta\Omega\, I_{\text{Acc}}}$	$\sigma_{\text{rel}}^{\text{finger}}$ (%)	$\frac{\Delta\Phi^{\text{bulk}}}{\Delta\Omega\, I_{\text{Acc}}}$	$\sigma_{\text{rel}}^{\text{bulk}}$ (%)	Gain factor
$r = 1\,\text{cm}, \varphi = 10°$	8.53×10^7	3.1	1.87×10^6	20.7	45.6
$r = 1\,\text{cm}, \varphi = 45°$	9.90×10^7	2.9	3.20×10^6	16.7	31.0
$r = 1\,\text{cm}, \varphi = 90°$	9.12×10^7	3.0	2.92×10^6	26.6	31.3
$r = 1\,\text{cm}, \varphi = 120°$	5.83×10^7	3.9	1.65×10^7	11.2	3.5
$r = 3\,\text{cm}, \varphi = 10°$	8.05×10^7	1.1	2.16×10^6	6.6	37.3
$r = 3\,\text{cm}, \varphi = 90°$	6.19×10^7	3.1	2.41×10^6	10.0	25.7

channels. To estimate the overall performance of the Finger Moderator the ratio of the peak brilliances for $\theta \leq 1°$ of the finger and the bulk moderator are considered which amounts to $\Delta\Phi_{\text{th}}^{\text{finger}}/\Delta\Phi_{\text{th}}^{\text{bulk}} = 45.6$ for the flux channel with $r = 1$ cm, $\varphi = 10°$ and $\Delta\Phi_{\text{th}}^{\text{finger}}/\Delta\Phi_{\text{th}}^{\text{bulk}} = 25.7$ for the flux channel with $r = 3$ cm, $\varphi = 90°$, see Table 4.8.

For an optimum performance of the instruments, care must be taken that the divergence, which can be transferred by a neutron guide or supermirror, respectively, is strongly energy-dependent. The data of Figs. 4.35 and 4.36 are therefore of limited applicability for the development of instruments because they are integrated over the thermal or cold energy range, respectively. Therefore, an extensive simulation study is performed in which the brilliance at the channel exits is determined as a function of the energy. The results for the previously discussed flux channels are shown in Figs. 4.37 and 4.38, respectively, whereas the results of the other channels is given in Figs. B.13–B.15 in the Appendix. The energy boundaries of the thermal and cold regimes are indicated by dotted lines. Integrating over the energy range between these boundaries yields the same values as given by the solid curves for 33 cm in Figs. 4.35 and 4.36.

By means of the existing data, the relevant area in the θ-E plane can now be integrated over the respective range in order to determine the phase space volume transported by the corresponding neutron guide.

4.2.5.3 Temporal Pulse Shape

The spectra in Fig. 4.32 and the energy-resolved brilliance in Figs. 4.37 and 4.38 clearly indicate that a distinct background of fast and epithermal neutrons is extracted through the flux channels. Due to the much different flight times of the neutrons at different energies, the background can be separated from the useful signal by choppers or time-resolved detectors.

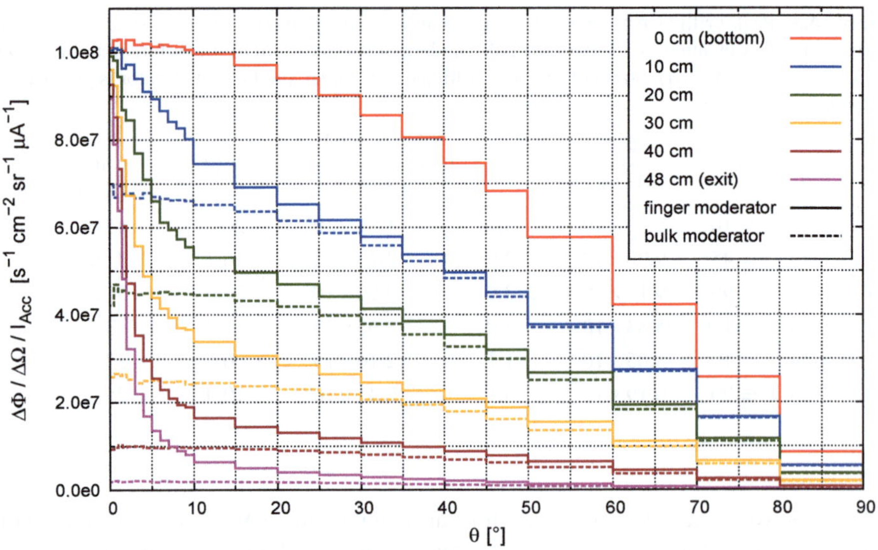

Fig. 4.35 Brilliance of thermal neutrons at different positions inside the flux channel with $r = 1\,\text{cm}$, $\varphi = 10°$ (solid lines) and a bulk moderator, for which the channel is filled with material up to the measuring position (dashed lines)

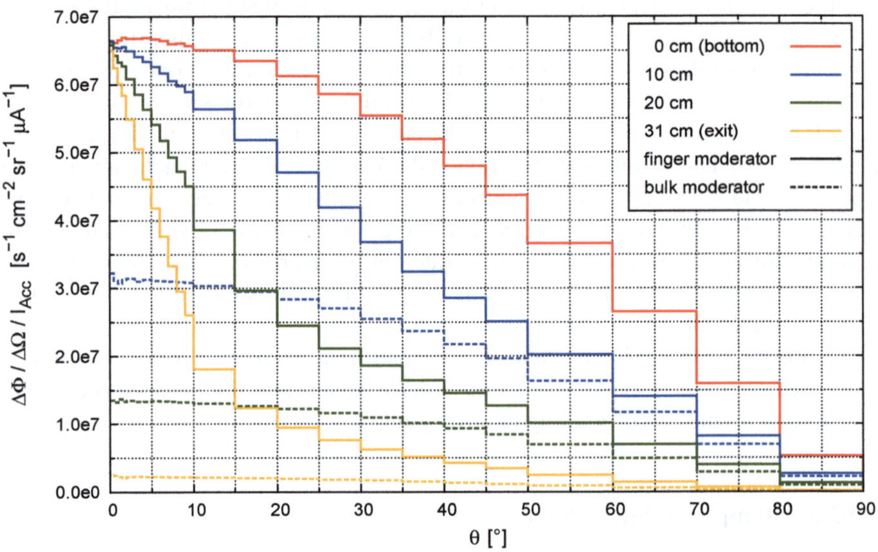

Fig. 4.36 Brilliance of thermal neutrons at different positions inside the flux channel with $r = 3\,\text{cm}$, $\varphi = 90°$ (solid lines) and a bulk moderator, for which the channel is filled with material up to the measuring position (dashed lines)

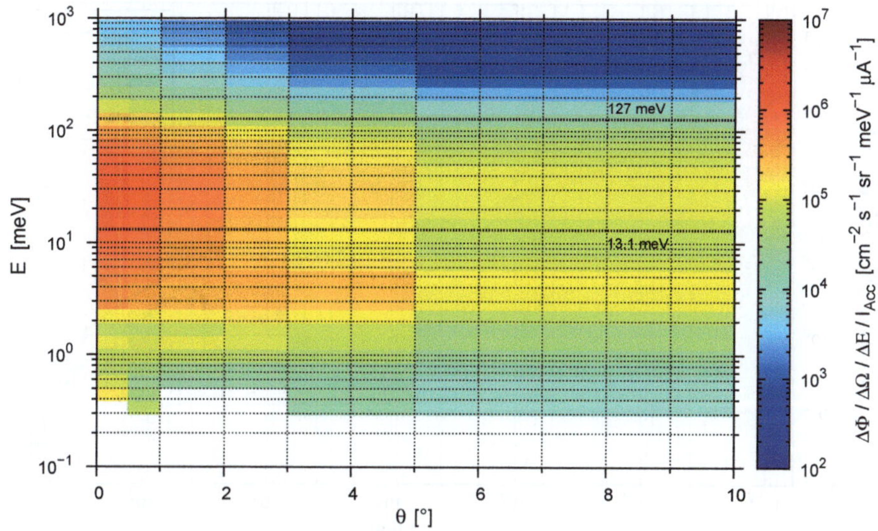

Fig. 4.37 Energy-resolved brilliance of neutrons leaving the channel with $r = 1$ cm, $\varphi = 10°$. The thermal regime is marked by horizontal dotted lines

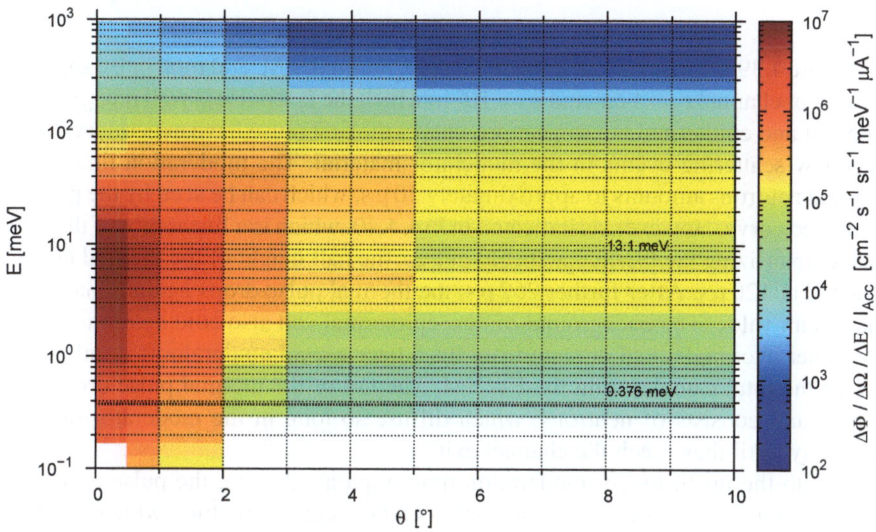

Fig. 4.38 Energy-resolved brilliance of neutrons leaving the parahydrogen channel ($r = 1$ cm, $\varphi = 45°$). The cold regime is marked by horizontal dotted lines

To support the development of instruments to this respect, the pulse structure at the channel exits in forward direction (1° with respect to the corresponding channel axis) is determined for cold, thermal and fast neutrons. In addition, the FWHM of the cold and thermal pulses is calculated and included in the plots.

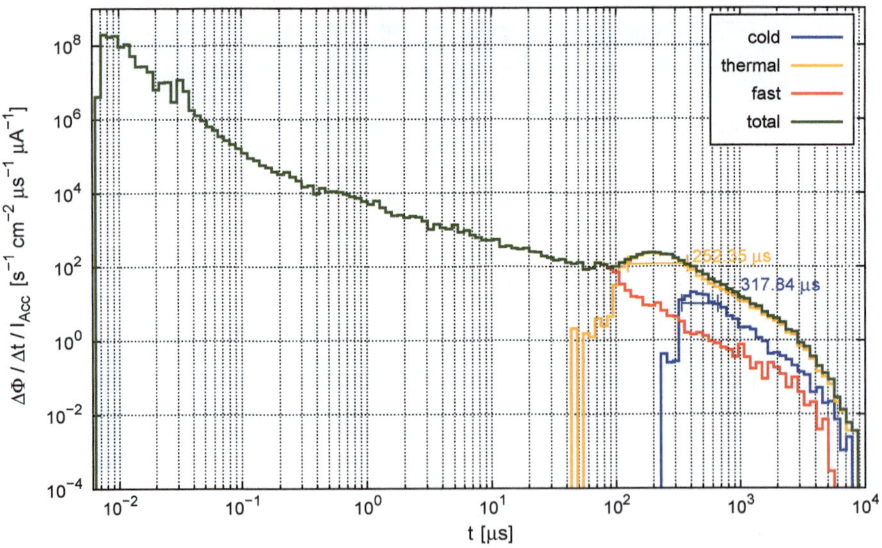

Fig. 4.39 Logarithmic plot of the pulse shape of cold, thermal and fast neutrons with a maximum divergence of $\theta = 1°$ in the channel with $r = 1$ cm, $\varphi = 10°$. The FWHM is indicated in the plot

Figure 4.39 is a logarithmic plot of the cold, thermal, fast, and total pulse shapes at the exit of channel $r = 1$ cm and $\varphi = 10°$ in which the fast pulse is well recognizable. The fast and epithermal neutrons representing the first pulse have experienced no or very few scattering events in the moderator material. The moderation time of the thermal neutrons amounts to approximately 50 µs, which can be seen by the position of the red curve. However, as indicated in Fig. 4.40, which is a linear plot of the same data emphasizing the thermal and cold pulses, the half height of the thermal pulse is reached at 120 µs. After further 252 µs, the thermal pulse drops again to half of its maximum value. The background of fast, epithermal and also cold neutrons (which are either overmoderated or stem from the adjacent cold moderators) is only about 5% of the total signal during the FWHM. The tail of the thermal pulse drops very slowly and consists of neutrons, which diffuse so long in the moderator-reflector assembly until they reach the channel exit.

Due to the much longer moderation time in parahydrogen, the pulse at the exit of the channel $r = 1$ cm and $\varphi = 45°$ shows a bi-spectral structure, which is clearly visible in the linear plot Fig. 4.41. The half maximum of the thermal pulse is reached at 190 µs and at 250 µs for the cold pulse. The FWHM of the thermal pulse amounts to only 143 µs but is still overlapping with the FWHM of the cold pulse. Due to this fact and the slowly decreasing tail of the thermal pulse, the cold pulse has a fairly large background. Due to the different flight times, however, the pulses can be separated with a sufficiently large flight path within the instrumentation. In experiments utilizing cold neutrons, the thermal pulse can be removed with a chopper.

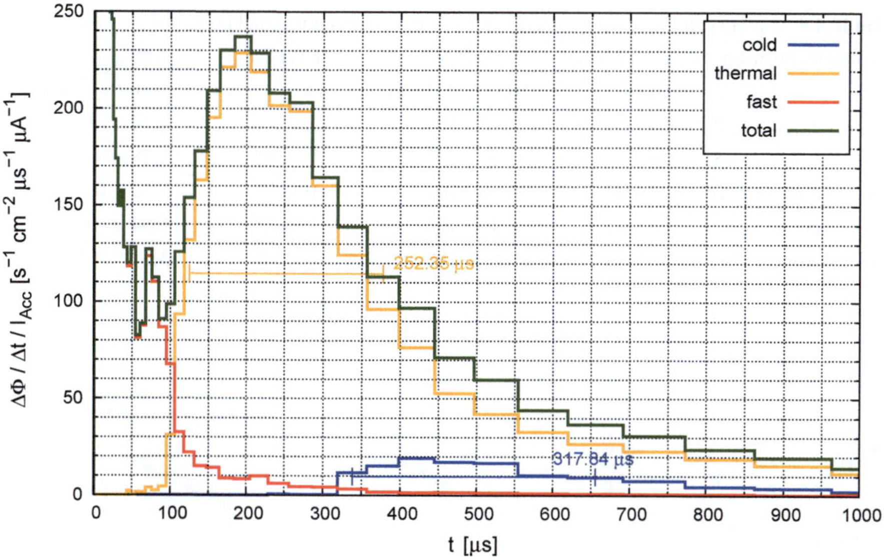

Fig. 4.40 Linear plot of the pulse shape of cold, thermal and fast neutrons with a maximum divergence of $\theta = 1°$ in the channel with $r = 1\,\text{cm}$, $\varphi = 10°$. The FWHM is indicated in the plot

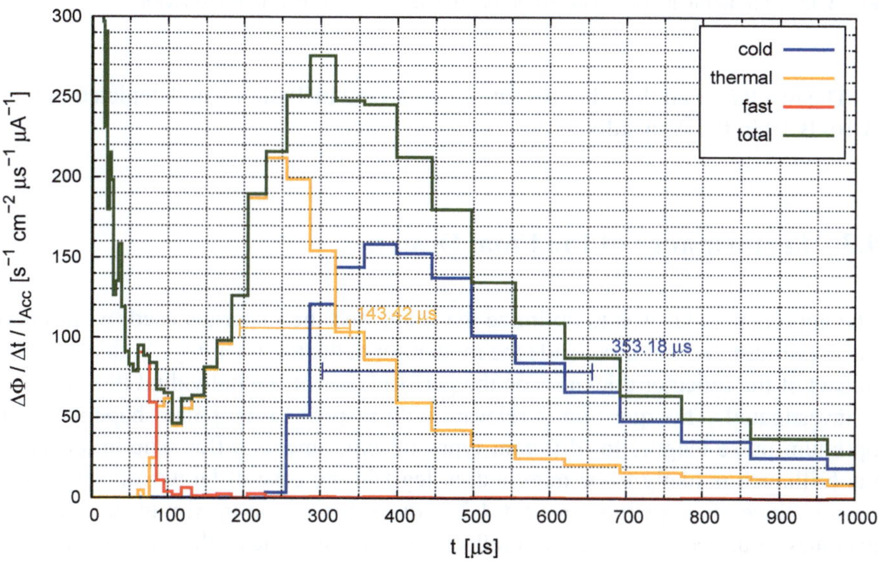

Fig. 4.41 Linear plot of the pulse shape of cold, thermal and fast neutrons with a maximum divergence of $\theta = 1°$ in the parahydrogen channel ($r = 1\,\text{cm}$, $\varphi = 45°$). The FWHM is indicated in the plot

Table 4.9 FWHM and flux during the FWHM of the cold and thermal pulses of all flux channels

Channel position	$FWHM_{cold}$	$\Delta\Phi^{cold}_{FWHM}$	$FWHM_{th}$	$\Delta\Phi^{th}_{FWHM}$
	(μs)	($cm^{-2}\,s^{-1}\,\mu A^{-1}$)	(μs)	($cm^{-2}\,s^{-1}\,\mu A^{-1}$)
$r = 1\,cm, \varphi = 10°$	317.34	5.15×10^3	252.35	4.74×10^4
$r = 1\,cm, \varphi = 45°$	353.18	4.64×10^4	143.42	2.78×10^4
$r = 1\,cm, \varphi = 90°$	178.90	4.86×10^3	190.52	5.59×10^4
$r = 3\,cm, \varphi = 10°$	642.89	3.43×10^4	321.22	1.54×10^4
$r = 3\,cm, \varphi = 90°$	95.33	2.01×10^3	162.18	3.51×10^4

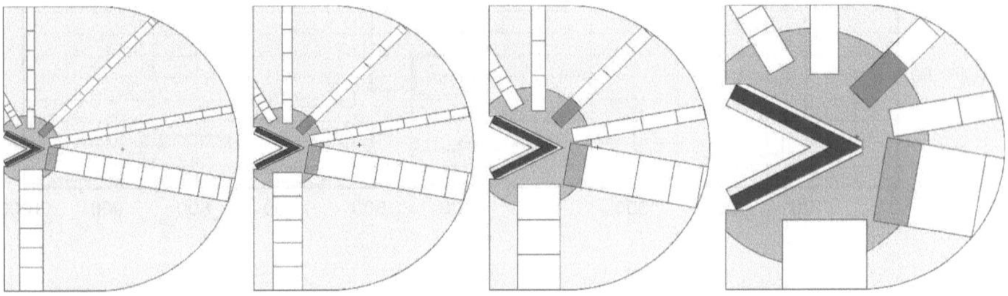

(a) Thickness 46 cm **(b)** Thickness 32 cm **(c)** Thickness 17 cm **(d)** Thickness 5 cm

Fig. 4.42 Simulation models with different thicknesses of the beryllium reflector

The results for all flux channels are summarized in Table 4.9 and depicted in Figs. B.16–B.19 in the Appendix.

4.3 Performance of the Final Design

The approach so far was to optimize the shape and the materials used by means of the maximum thermal flux inside the moderator-reflector assembly. At this point the impact of the reflector thickness on the performance of the individual flux channels is investigated. This question is relevant with respect to the pulse shape since the optimum pulse lengths for different types of experiments differ. The influence on the brilliance is likely to be marginal, at least in the case of larger reflector thicknesses, since, as explained in Sect. 4.2.3, the value of the maximum thermal flux becomes saturated above a certain reflector volume and the brilliance in the forward direction is extracted by the flux channels from the location of the flux maximum of the moderator-reflector assembly. Four different models are examined which differ in the reflector thickness having the values 46, 32, 17, and 5 cm. These values refer to the thickness of the beryllium reflector from the surface of the PE sphere in the z-direction, e.g., the axis of symmetry of the moderator-reflector assembly (Fig. 4.42).

Another circumstance which has not yet been considered is the influence of the finite pulse length of the accelerator. The previous results all relate to an instantaneous bombardment of the target by the initial protons. As mentioned in Sect. 4.1.3, an accelerator is considered which provides an average power of 100 kW with an energy of 50 MeV. The duty cycle D of the accelerator is calculated as follows:

$$\begin{aligned} P_{\text{peak}} &= I_{\text{peak}} \cdot E = 100\,\text{mA} \cdot 50\,\text{MeV} = 5\,\text{MW} \\ \overline{P} &\overset{!}{=} 100\,\text{kW} = P_{\text{peak}}/50 \\ \Rightarrow D &= 1:50 \end{aligned}$$

A pulse length of $\tau_{\text{acc}} = 200\,\mu\text{s}$ is obtained for a repetition rate of $f = 100\,\text{Hz}$ together with the duty cycle $D = 1{:}50$. With this pulse length, extensive simulations are carried out in which the brilliance and the pulse structure of the neutrons are determined at the channel exits for different reflector thicknesses. For this purpose a new source file has to be created, according to Sect. 4.1.1, to account for the finite pulse width of the accelerator. The results of the simulations are discussed hereafter.

4.3.1 Brilliance

Figure 4.43 shows the thermal brilliance of the channel with $r = 1\,\text{cm}$, $\varphi = 10°$ for all examined reflector thicknesses. As previously assumed, the dependence of the peak brilliance on the reflector thickness is relatively weak, except for the smallest thickness examined. The highest peak brilliance is achieved with a reflector thickness of 32 cm resulting from the geometrical effect that not only neutrons from the bottom of the channel contribute to the peak brilliance, but also neutrons emitted at very acute angles from the inner walls of the channels. However, the permissible angles, for which neutrons contribute to the peak brilliance, of the lower regions of the inner walls become smaller with increasing reflector thicknesses. This effect competes with the fact that the more neutrons are emitted from the channel bottom the more reflector material is used. Thus, there is a finite optimum reflector thickness. However, the divergence of the neutrons passing through the channel exit is significantly increasing for decreasing reflector thickness. Hence, the shape of the brilliance becomes increasingly flatter. This effect is of geometric nature as explained in Sect. 4.2.5.2.

As can be seen in Fig. 4.44, the above statements can be qualitatively transferred to the channel with $r = 1\,\text{cm}$, $\varphi = 45°$ with parahydrogen. However, the cold neutrons, which, apart from overmoderated neutrons from the beryllium reflector, are emitted exclusively from the outer face of the cold moderator, i.e., the channel floor. In the first angle bin of 0° to 0.5°, the brilliance is therefore the highest for the thickest reflector. However, it should be noted that neutron guides for cold neutrons accept a significantly larger divergence of up to 4°, as explained in Sect. 2.4. In this respect the reflector with 17 cm delivers the highest neutron flux as indicated in Table 4.10.

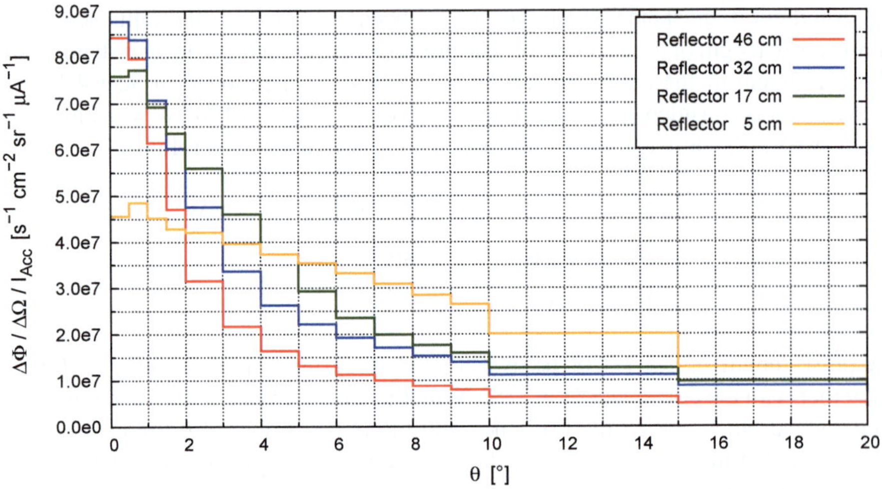

Fig. 4.43 Brilliance of thermal neutrons leaving the channel with $r = 1\,\text{cm}$, $\varphi = 10°$ for all reflector thicknesses investigated and a proton pulse width of 200 μs

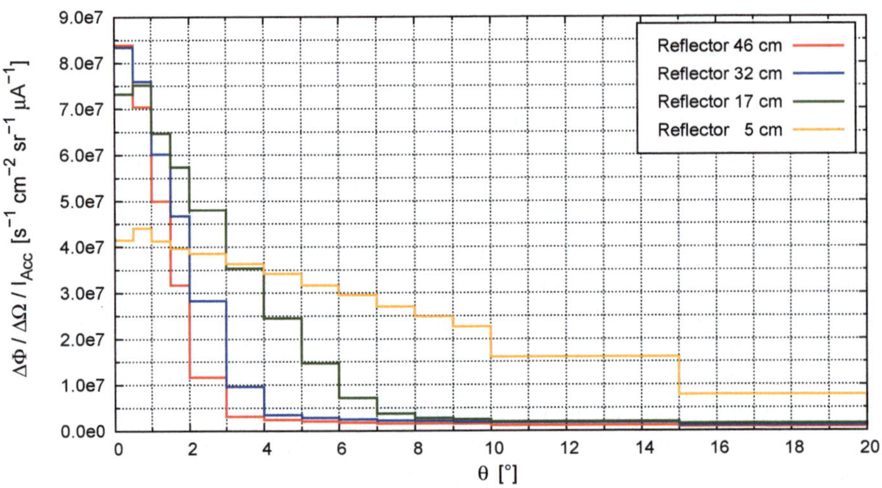

Fig. 4.44 Brilliance of cold neutrons leaving the parahydrogen channel with $r = 1\,\text{cm}$, $\varphi = 45°$ for all reflector thicknesses investigated and a proton pulse width of 200 μs

However, for the larger solid methane moderator the optimum reflector thickness amounts to 32 cm as shown in Fig. B.22 in the Appendix. The results of the other flux channels is given in Figs. B.20–B.23.

The pulse structure of the accelerator has no influence on the results shown here since the brilliance is not determined time-resolved.

Table 4.10 Cold flux of the parahydrogen channel integrated over the first 4° which is a typical value of the critical angle of supermirrors for cold neutrons

Thickness (cm)	$\Delta\Phi^{\text{cold}}(\theta < 4°)$ ($\text{cm}^{-2}\,\text{s}^{-1}\,\mu\text{A}^{-1}$)	σ_{rel} (%)
46	8.13×10^5	1.25
32	1.33×10^6	0.98
17	2.23×10^6	0.51
5	1.84×10^6	0.55

4.3.2 Temporal Pulse Shape

The results for the channel with $r = 1$ cm, $\varphi = 10°$ and $r = 1$ cm, $\varphi = 45°$ are discussed exemplarily. The pulse structure of the fast, thermal, and cold neutrons of the former is shown for a reflector thickness of 46 cm in Fig. 4.45.

In comparison to the pulse shapes for instantaneous irradiation of the target as discussed in Sect. 4.2.5.3, it is apparent that the flux of fast neutrons is nearly constant during the 200 μs proton pulse of the accelerator. The FWHM of the thermal pulse is reached after 220 μs, i.e., 100 μs later than for instantaneous irradiation, cf. Fig. 4.40. Even taking into account the quite long proton pulse, the FWHM of the thermal pulse, which amounts to 277 μs, is not much longer than for instantaneous irradiation.

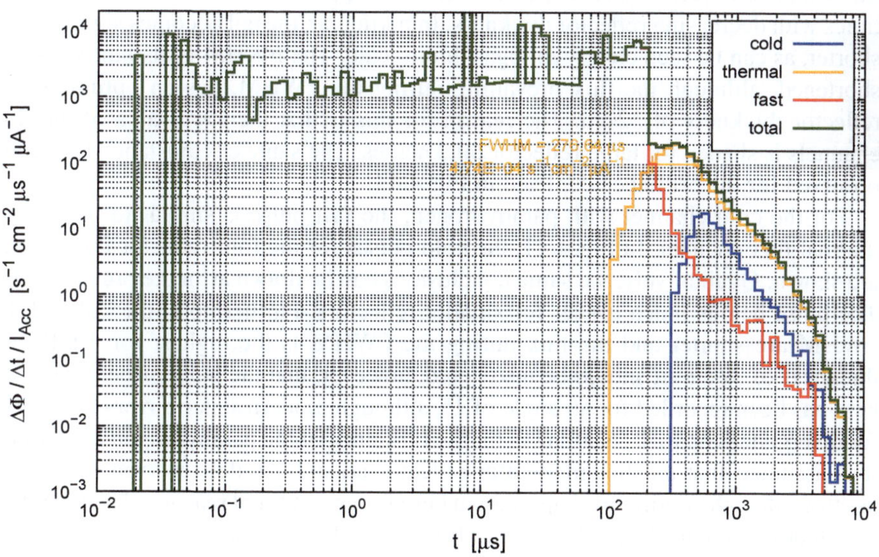

Fig. 4.45 Pulse shape of cold, thermal and fast neutrons in the channel with $r = 1$ cm, $\varphi = 10°$ for a reflector thicknesses of 46 cm in a maximum divergence of $\theta = 1°$ for a proton pulse width of 200 μs. FWHM and thermal flux during the FWHM are stated in the plot

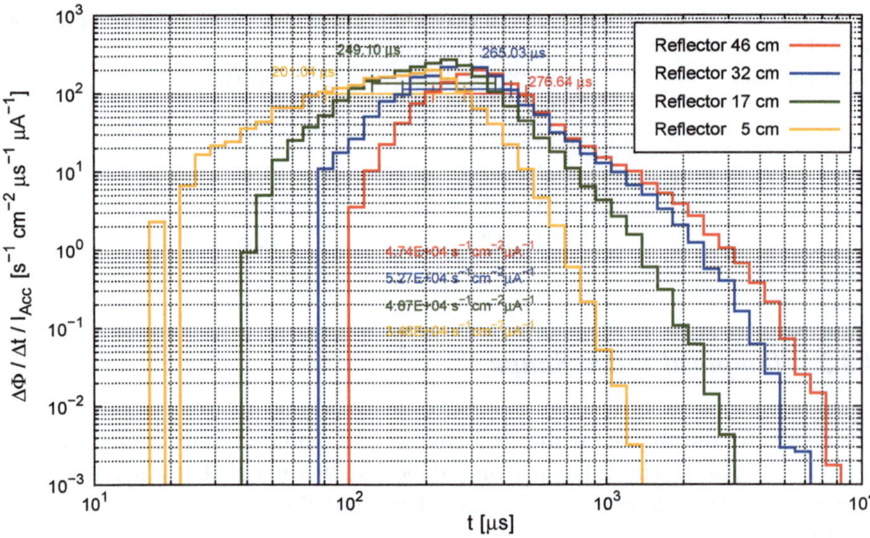

Fig. 4.46 Pulse shape of thermal neutrons leaving the channel with $r = 1\,\text{cm}$, $\varphi = 10°$ for all reflector thicknesses investigated in a maximum divergence of $\theta = 1°$ for a proton pulse width of $200\,\mu\text{s}$. FWHM and thermal flux during the FWHM are stated in the plot

In Fig. 4.46, the thermal pulses for all reflector thicknesses are combined for the flux channel in question. As expected, the thermal pulse continues to shift to earlier times with decreasing reflector thickness. In particular, the tail becomes significantly shorter, as can be seen clearly in logarithmic scale. The FWHM is also increasingly shortened, although the flux prevailing during the FWHM, has an optimum at a reflector thickness of 32 cm. The reason for this is that the lower area of the flux channels is shaded for large reflector thicknesses. The same circumstance is already observed for the brilliance depicted in Fig. 4.43.

For the cold channel with parahydrogen, the fast pulse is similar to that in the thermal flux channel. In Fig. 4.47, the cold pulses for the four reflector thicknesses are shown. Qualitatively, the same conclusions can be drawn as in the previous part. The pulse begins much earlier for smaller reflectors. Furthermore, the tail and the FWHM are shorter and the flux during the FWHM is the highest for a reflector thickness of 32 cm. It should be also noted, however, that the pulse structure shown here has been recorded for an angle of 1°, although modern supermirrors also allow for divergences up to 4°. Similar as for the brilliance of the parahydrogen moderator, it is assumed that smaller moderator-reflector assemblies deliver even more cold neutrons than the optimum shape shown here for geometrical reasons. The results of the other flux channels are depicted in Figs. B.24–B.27 in the Appendix.

For designing specific experiments and instruments, it must also be considered that the flight times for neutrons, which have thus far been summarized as thermal or cold, differ so strongly that in some types of experiments not the entire spectrum

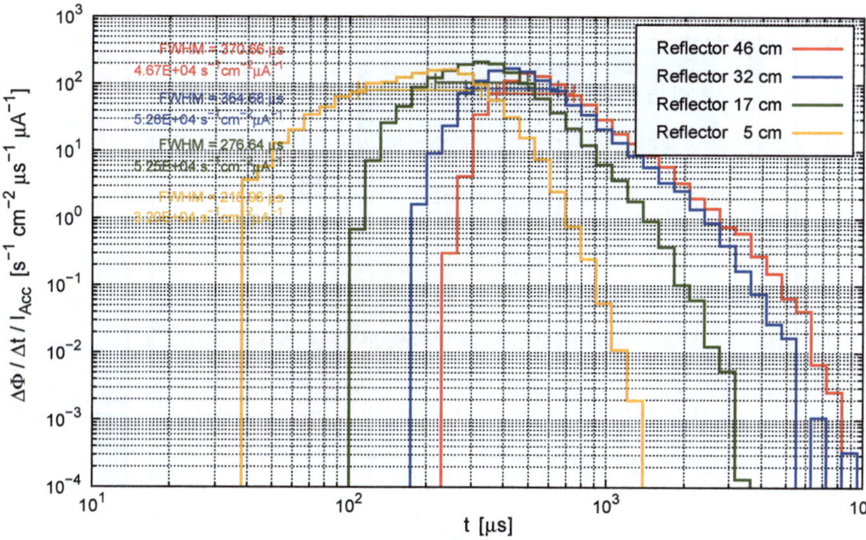

Fig. 4.47 Pulse shape of cold neutrons leaving the parahydrogen channel with $r = 1$ cm, $\varphi = 45°$ for all reflector thicknesses investigated in a maximum divergence of $\theta = 1°$ for a proton pulse width of 200 μs. FWHM and cold flux during the FWHM are stated in the plot

of thermal or cold neutrons can be used over the full pulse duration. In order to be able to estimate the performance of those instruments, the pulse shapes are also determined energy-resolved. In this case, the permissible opening angle is set to 5°, since otherwise the statistical error would become too large. The thermal and cold energy regimes are bordered by dashed lines, the end of the proton pulse is marked by a horizontal solid line.

Results for the thermal flux channel for reflector thicknesses 46 and 5 cm are depicted in Fig. 4.48a, b, respectively. It is very striking that the primary neutrons are moderated very quickly after the proton pulse. As described earlier in this section, the moderation time is approximately 50 μs, so that the flux density of fast neutrons already drops significantly in the first time bin after the proton pulse (200–259 μs). At even later times, the background to the thermal pulse, apart from overmoderated cold neutrons, consists only of epithermal neutrons with a maximum of 475 meV. Furthermore it is apparent, that the maximum values of the thermal pulse are larger for smaller reflector thicknesses. This can be explained by the fact that the results shown here were taken for a divergence of 5°, whereas the previously discussed energy-independent pulse structures apply for a divergence of 1°. For geometrical reasons, the brilliance increases with decreasing reflector thickness, which is also evident by considering the flatter brilliance in Fig. 4.43. Since typical neutron guides can transport a divergence of a maximum of 1° for thermal neutrons, these results should only be used if the exact energy dependence of the pulse structure has to be considered. A reflector with a thickness of 32 cm offers the best performance for

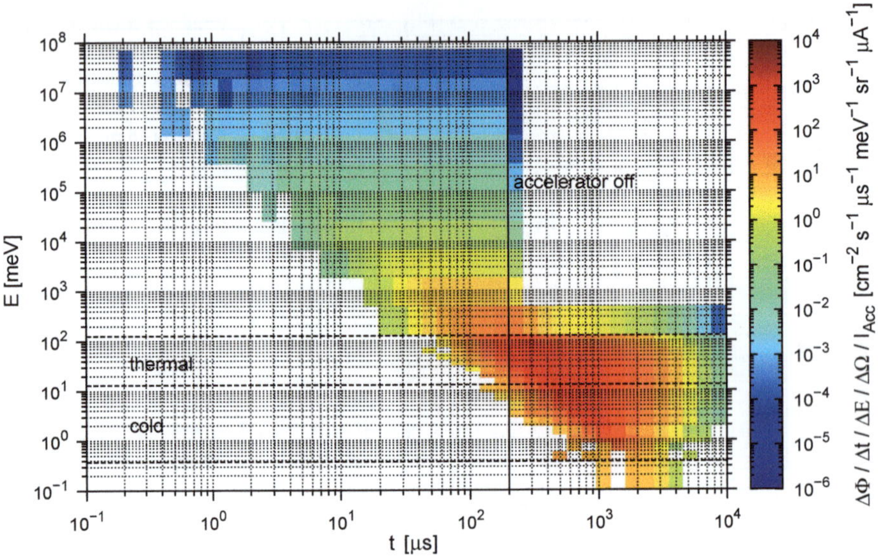

(a) Thickness 46 cm.

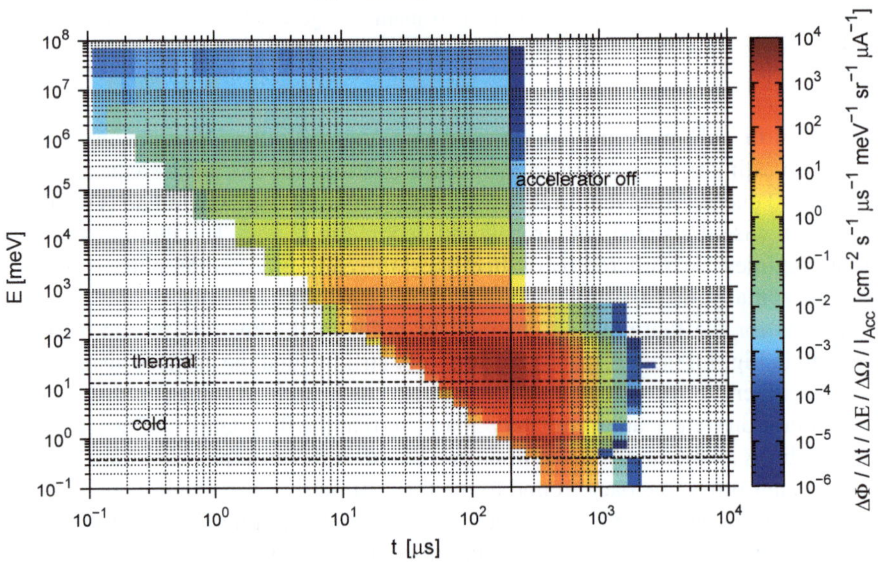

(b) Thickness 5 cm.

Fig. 4.48 Energy-resolved pulse shape of the thermal flux channel ($r = 1\,\text{cm}$, $\varphi = 10°$) for different reflector thicknesses. Neutrons with a divergence of $\theta \leq 5°$ are considered for better statistics

thermal flux channels according to Figs. 4.43 and 4.46. The energy-resolved pulse structures shown here should be best used in conjunction with the brilliance from Fig. 4.43, for example by normalizing by the integral flux values from 0° to 1°. This

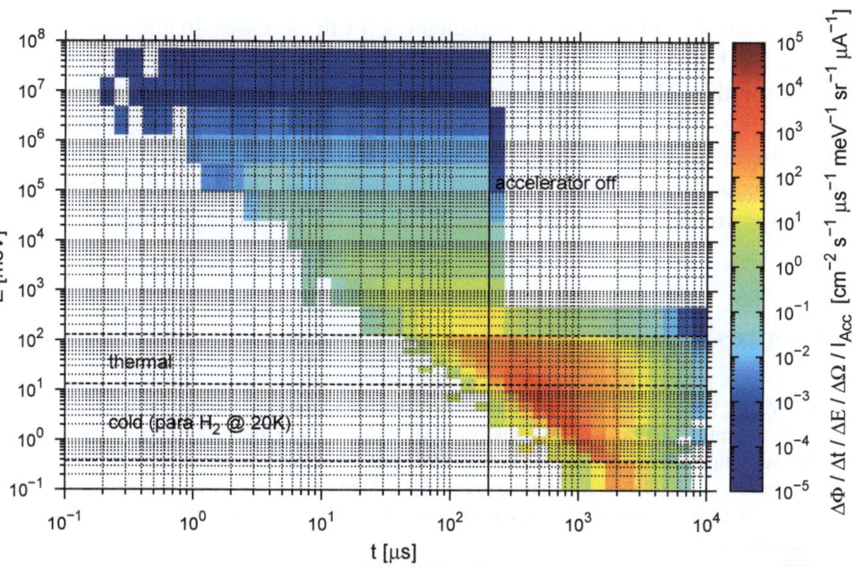

(a) Thickness 46 cm.

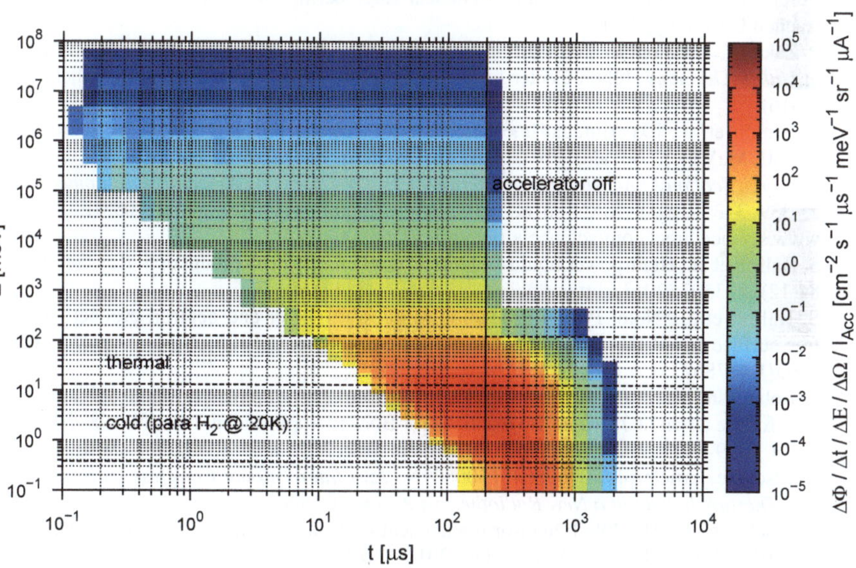

(b) Thickness 5 cm.

Fig. 4.49 Energy-resolved pulse shape of the parahydrogen flux channel ($r = 1\,\text{cm}$, $\varphi = 45°$) for different reflector thicknesses. Neutrons with a divergence of $\theta \leq 5°$ are considered for better statistics

procedure is allowed provided that the pulse forms and the spectra do not differ greatly between 1° and 5°.

Results for the cold parahydrogen channel for thicknesses 46 and 5 cm are given in Fig. 4.49a, b, respectively. As discussed earlier, the same effect comes into play for what reason the cold pulses are higher and shorter for smaller reflectors and larger divergences, see Sect. 4.3.1. Since neutron guides accept a divergence of a maximum of 4°, these diagrams give a better estimate of the performance of the cold moderator than the plots for the thermal channel. The results of the other flux channels for thicknesses 46 and 5 cm are given in Figs. B.28–B.31 in the Appendix.

References

1. X-5 Monte Carlo Team, MCNP-A General N-Particle Transport Code, Version 5. LA-UR-03-1987. Technical report Los Alamos National Laboratory, 1 Feb 2008. https://laws.lanl.gov/vhosts/mcnp.lanl.gov/pdf_files/la-ur-03-1987.pdf
2. A. Sperling, J.P. Dabruck, binning.java., *Java Application for Twodimensional Binning of Scattered Data* (Institute of Nuclear Engineering and Technology Transfer, RWTH Aachen University, 2015)
3. A. Nalbandyan, *Simulation of Total Neutron Yields Produced by Proton Beams on Be Target* (Report on Mini-Project. Institute of Nuclear Engineering and Technology Transfer, RWTH Aachen University, 2016)
4. P. Zakalek, Development of high-brilliant neutron source targets, in *Workshop Presentation in Unkel, Germany*. Jülich Center for Neutron Science 2, Forschungszentrum Jülich, vol. 29 (2016)
5. I. Tilquin et al., Experimental measurements of neutron fluxes produced by proton beams (23-80 MeV) on Be and Pb targets, in: *Nuclear Instruments and Methods in Physics Research Section A: Accelerators, Spectrometers, Detectors and Associated Equipment* 545.1–2, 339–343 (2005). ISSN: 0168-9002. https://doi.org/10.1016/j.nima.2005.01.325, http://www.sciencedirect.com/science/article/pii/S016890020500522X
6. S. Halfon et al., High-power liquid-lithium jet target for neutron production. Rev. Sci. Instrum. **84**(12), 123507 (2013). https://doi.org/10.1063/1.4847158
7. S. Halfon et al., Note: Proton irradiation at kilowatt-power and neutron production from a free-surface liquid-lithium target. Rev. Sci. Instrum. **85.5**, 056105 (2014). https://doi.org/10.1063/1.4878627
8. T. Kobayashi, K. Miura, N. Hayashizaki, M. Aritomi, Development of liquid-lithium film jet-flow for the target of 7Li(p,n)7Be reactions for BNCT. Appl. Radiat. Isot. **88**, 198–202 (2014). ISSN 0969-8043. https://doi.org/10.1016/j.apradiso.2013.12.013, http://www.sciencedirect.com/science/article/pii/S096980431300609X (*15th International Congress on Neutron Capture Therapy Impact of a New Radiotherapy Against Cancer*)
9. U. Rücker et al., The Jülich high-brilliance neutron source project. Eur. Phys. J. Plus **131**(1), 19 (2016). https://doi.org/10.1140/epjp/i2016-16019-5
10. J. Ulrich, *Untersuchungen zurNeutronenproduktion mittels photonuklearer Wechselwirkungen* (Available for internal use at NET, RWTH Aachen and its project partners, Aachen, Germany, 31 Dec 2015)
11. F.J. Bermejo, F. Sordo, ESS-Bilbao Target Station. Technical Design Report. ESS-Bilbao Target Devision, 1 July 2013, p. 33. ISRN: 978-84-695-8105-6. http://www.essbilbao.org/index.php/en/component/docman/doc_download/417-essb-ntar-2013-final-tdr-july
12. L.D. Landau, E.M. Lifschitz, *Lehrbuch der theoretischen Physik Band 1: Mechanik* (Akademie Verlag, Berlin, Germany, 1970). ISBN: 978-3808556122
13. D. Meschede, *Gerthsen Physik* (Springer, Berlin, 2010). ISBN: 978-3-642-12893-6

14. Y. Yamagata, J. Ju, K. Hirota, Neutron generation source, and neutron generation device. EP Patent App. EP20130757266 (Japan) (2015). https://patents.google.com/patent/EP2824999A1/ja
15. H. Kumada et al., Development of beryllium-based neutron target system with three-layer structure for accelerator-based neutron source for boron neutron capture therapy. Appl. Radiat. Isot. **106**, 78–83 (2015). https://doi.org/10.1016/j.apradiso.2015.07.033, http://www.sciencedirect.com/science/article/pii/S0969804315301159 (*The 16th International Congress on Neutron Capture Therapy (ICNCT-16). Representative person of the Organizing Committee: Dr Hanna Koivunoro (Secretary general of the ICNCT-16)*)
16. T. Rinckel, D.V. Baxter, J. Doskow, P.E. Sokol, T. Todd, Target performance at the low energy neutron source. Phys. Proc. **26**, 168–177 (2012). ISSN: 1875-3892. https://doi.org/10.1016/j.phpro.2012.03.022, http://www.sciencedirect.com/science/article/pii/S1875389212004385
17. H. Kumada et al., Development of beryllium-based neutron target system with three-layer structure for accelerator-based neutron source for boron neutron capture therapy. Appl. Radiat. Isot. **106**, 78–83 (2015). ISSN: 0969-8043. https://doi.org/10.1016/j.apradiso.2015.07.033, http://www.sciencedirect.com/science/article/pii/S0969804315301159 (*The 16th International Congress on Neutron Capture Therapy (ICNCT-16). Representative person of the Organizing Committee: Dr Hanna Koivunoro (Secretary general of the ICNCT-16)*)
18. J.M. Carpenter, Gallium-cooled target for compact accelerator-based neutron sources. Phys. Proc. **26**, 132–141 (2012). ISSN: 1875-3892. https://doi.org/10.1016/j.phpro.2012.03.018, http://www.sciencedirect.com/science/article/pii/S1875389212004348
19. J.W. Westwater, L.S. Tong, Boiling heat transfer and two-phase flow. AIChE J. **12.3**, 616–617 (1966). ISSN: 1547-5905. https://doi.org/10.1002/aic.690120303 (Wiley)
20. M.M. El-Wakil, *Nuclear Heat Transport* (American Nuclear Society, 1978). ISBN: 978-0-89448-014-0
21. B.W. Blackburn, J.C. Yanch, R.E. Klinkowstein, Development of a high power water cooled beryllium target for use in accelerator-based boron neutron capture therapy. Med. Phys. **25**(10), 1967–1974 (1998). ISSN: 2473- 4209. https://doi.org/10.1118/1.598370
22. I. Silverman, A. Nagler, High heat flux cooling with water jet impingement, in *ASME 2004 Heat Transfer/Fluids Engineering Summer Conference*, vol. 1 (2004), pp. 277–282. https://doi.org/10.1115/HT-FED2004-56273
23. J. Esposito et al., Be target development for the accelerator-based SPES-BNCT facility at INFN Legnaro. Appl. Radiat. Isot. **67**, 7–8 (2009). ISSN: 0969-8043. https://doi.org/10.1016/j.apradiso.2009.03.085, http://www.sciencedirect.com/science/article/pii/S0969804309003017 (*13th International Congress on Neutron Capture Therapy BNCT: a new option against cancer, S270–S273*)
24. C. Ceballos et al., Towards the final BSA modeling for the accelerator-driven BNCT facility at INFN LNL. Appl. Radiat. Isot. **69**(12), 1660–1663 (2011). ISSN: 0969-8043. https://doi.org/10.1016/j.apradiso.2011.01.032. http://www.sciencedirect.com/science/article/pii/S0969804311000467 (*Special Issue: 14th International Conference on Neutron Capture Therapy*)
25. B.W. Blackburn, J.C. Yanch, *Liquid Gallium Cooling for a High-Power Beryllium Target for use in Accelerator Boron Neutron Capture Therapy (ABNCT)* (Target Chemistry, 1999)
26. G. Speckbrock, S. Kamitz, M. Alt, H. Schmitt, *Clinical Thermometer*. EP0657023 (1996). http://www.freepatentsonline.com/EP0657023B1.html(visitedon09/11/2016)
27. M. Winter, *Gallium: The Essentials* (The University of Sheffield and Web Elements Ltd., 1993–2016). https://www.webelements.com/gallium/
28. Q. Xu, N. Qudalov, Q. Guo, H. Jaeger, E. Brown, *Effect of Oxidation on the Mechanical Properties of Liquid Gallium and Eutectic Gallium-Indium* (2012), p. 18. https://doi.org/10.1063/1.4724313, https://arxiv.org/abs/1201.4828v1
29. Geratherm Medical AG, Galinstan fluid. Safety Data Sheet. Germany, 18 Mar 2004. http://www.rgmd.com/msds/msds.pdf(visitedon09/11/2016)
30. M.D. Dickey et al., Eutectic Gallium-Indium (EGaIn): a liquid metal alloy for the formation of stable structures in microchannels at room temperature. Adv. Funct. Mater. **18**(7), 1097–1104 (2008). ISSN: 1616-3028. https://doi.org/10.1002/adfm.200701216

31. Y. Zhang, J.R.G. Evans, S. Yang, Corrected values for boiling points and enthalpies of vaporization of elements in handbooks. J. Chem. Eng. Data **56**(2), 328–337 (2011). https://doi.org/10.1021/je1011086
32. U. Rücker, T. Brückel, T. Cronert, J.P. Dabruck, R. Nabbi, Vorrichtung zur Erzeugung von thermischen Neutronenstrahlen mit Hoher Billanz und Herstellungsverfahren. PUB:(DE-HGF)23 Patent 16166567.4-1556 (EP3091540B1) (Germany). JCNS-2/PGI-4/JARA-FIT (2016). http://juser.fz-juelich.de/record/810621
33. R. Kajimoto et al., High intensity chopper spectrometer 4SEASONS At J-PARC. J. Neutron Res. **15**(1), 5–12 (2007). https://doi.org/10.1080/10238160601048742
34. K. Nakajima et al., AMATERAS: a cold-neutron disk chopper spectrometer. J. Phys. Soc. Jpn. **80**(Suppl.B), SB028 (2011). https://doi.org/10.1143/JPSJS.80SB.SB028

Chapter 5
Prototype Moderator at the AKR-2 Training Reactor

The simulation results on the optimization studies of a laser- and an accelerator-driven neutron source which are shown in Chaps. 3 and 4, respectively, are to be verified by means of experimental measurements on a prototype moderator. For this purpose, it is advisable to test the moderator at an existing and well-characterized neutron source with the advantage that additional effects like the ion-target interaction can be neglected.

A well-suited neutron source is the training and research reactor AKR-2 (Ausbildungskernreaktor Dresden) of the department for Hydrogen and Nuclear Energy Technology of the Dresden Technical University. The first criticality of the reactor took place on March 22, 2005. It is mainly used for training and teaching purposes, but also offers some experimental channels for conducting various research projects. The maximum power of the reactor is 2 W and is therefore referred to as a zero-power reactor. Since materials for cold moderators are to be investigated at the AKR-2, the low power is very advantageous because there is virtually no heat input into the moderator. The experimental setup is therefore less complex, whereby the neutron-physical results of the measurements can be easily scaled up. In addition, a very low activation of the measurement structures is to be expected due to the relatively low neutron flux, which is also preferable from a radiation protection point of view.

5.1 Design and Operational Features of the Reactor

The cylindrical reactor core exhibits a diameter of 250 mm and a critical height of 275 mm consisting of plate-shaped fuel elements with a total mass of 794 g of ^{235}U. The fuel element plates consist of a homogeneous mixture of uranium oxide with an enrichment of 19.8% ^{235}U and polyethylene as a moderator material. The maximum continuous power of the reactor is 2 W. The system is controlled by three combined control and safety rods made of cadmium. The neutron source for initiating the chain

J. P. Dabruck, *Target Station Optimization for the High-Brilliance Neutron Source HBS*, Springer Theses, https://doi.org/10.1007/978-3-030-05639-1_5

(a) Picture of the AKR-2 with measuring instruments for size comparison.

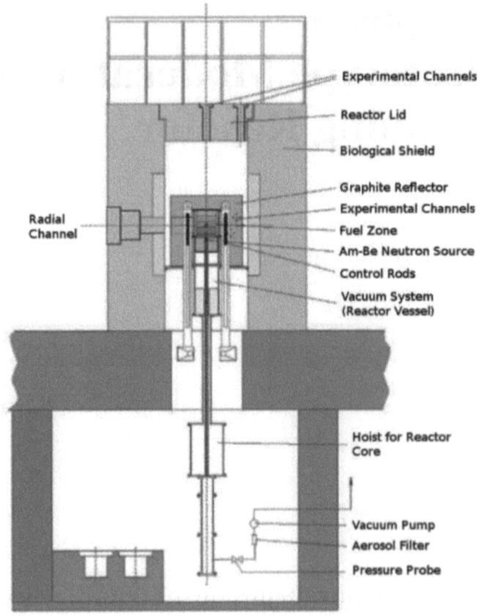

(b) Vertical cut of the constructions diagram of the AKR-2.

Fig. 5.1 The AKR-2 [1, 2]

reaction (Am-Be, source strength 2.2×10^6 n/s) is located inside the reactor. The core is surrounded by a reflector of pure graphite on all sides. The cylindrical reactor vessel is made of heavy concrete with a wall thickness of about 75 cm and serves as a radiation shielding.

For carrying out experiments, the AKR has the central channel leading through the reactor core as well as five further experimental channels. A larger experimental channel with a stepped cross section leads radially to the outer side of the reactor tank. This channel has a diameter of 200 mm directly at the reactor tank and a diameter of 500 mm at the exit. At a reactor power of 2 W, the flux density of thermal neutrons in the center of the central experimental channel amounts to approximately $2.5 \times 10^{-3}\,\mathrm{cm}^{-2}\,\mathrm{s}^{-1}$. Two ducts with a diameter of 75 mm are located tangentially to the core. The overall height of the apparatus amounts to 270 cm with a diameter of 340 cm. A picture and a schematic representation can be seen in Fig. 5.1.

In accordance with safety requirements the reactor is designed in such a way that a prompt criticality cannot be caused by maloperation under any circumstances and thus hazard for the operating personnel, the environment, or the reactor is impossible. For this aim, the excess reactivity is limited to a maximum of 0.3% and the release of fission products is prevented by a multi-barrier system [1, 2].

5.2 Characterization of the Experiments

By means of the experimental measurements on the prototype, various results of the simulation studies are to be examined. This includes the effect of the thermal flux channel, which can significantly increase the peak brilliance (see Sects. 3.5.1 and 4.2.5.2). For this purpose, not only the brilliance at the channel exit should be measured but also the brilliance at an equivalent area of the surface of the bulk moderator. In this way, the factor can be determined by which the peak brilliance can be increased by using a flux channel.

Another important point is the investigation of the performance of different cold moderators. These include hydrogen (ortho and para), solid methane and mesitylene. As discussed in Sect. 2.3.2, there are inconsistencies regarding the scattering cross section of parahydrogen which are to be tested at the AKR-2.

In addition, hydrogen moderators with different mixing ratios of ortho- and parahydrogen will be studied aiming for the determination of the most efficient ratio. As the scattering cross section of parahydrogen below 50 meV decreases significantly (see Fig. 2.8a, b), the generated cold neutrons are not in thermal equilibrium with the moderator and therefore undermoderated, which is evident by considering the corresponding spectra in Fig. 3.14. An admixture of orthohydrogen could cause the scattering cross section of the cold moderator to be increased so far that the neutrons can be brought into thermal equilibrium without having to increase the volume noticeably. To this end, a cryo system is developed in cooperation with TU Dresden, Chair for Hydrogen and Nuclear Energy Technology, in which the mixing ratio of ortho- and parahydrogen can be adjusted during the measurement and can be precisely determined by heat capacity measurement [3, 4].

5.3 The MCNP Reactor Model

The scientific staff of the AKR-2 provided an MCNP model [5] whose geometry is depicted in Fig. 5.2 as a vertical and a horizontal cut. The two core halves consisting of the fuel-moderator mixture are shown in magenta in the simulation model and are encompassed by a stainless steel housing (yellow) whereby the area between the housing an the reactor core is lined with graphite acting as a reflector (green). A large amount of additional graphite is located around the steel housing which is in turn surrounded by a biological shielding consisting of heavy concrete (salmon), with a ring of paraffin (orange) in the inner area at the level of the graphite reflector to slow down fast neutrons. An air gap (blue) extends between the reflector and the shielding and the entire shielding is bordered by an iron jacket (cyan) both inside and outside. The model does not include the upper shielding cover, the vacuum system, the neutron source, and all installations below the floor. The calculations are carried out in the `KCODE` criticality mode [6] initiated by neutrons defined with `SDEF` according to a Maxwell spectrum [5]. The validation of the model is described

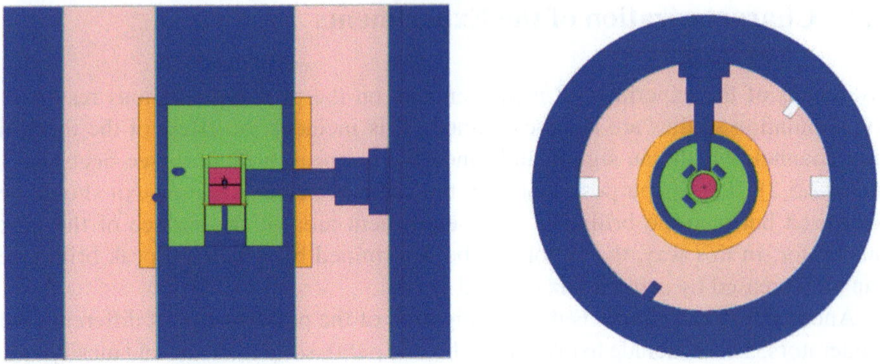

(a) Vertical cut of the MCNP geometry.

(b) Horizontal cut of the MCNP geometry at the height of the upper core half.

Fig. 5.2 MCNP Model of the AKR-2. The main materials defined in the model are fuel (magenta), stainless steel (yellow), graphite (green), heavy concrete (salmon), paraffin (orange), air (blue), and iron (cyan)

in the next section. On the basis of this, the correct normalization is derived, which can also be calculated analytically from the fission rate R_{fission} for a thermal power of 2 W by assuming a mean fission energy release of 180 MeV[1] and the mean number of fission neutrons of 2.44 in ^{235}U:

$$\begin{aligned} R_{\text{fission}} &= 2\,\frac{\text{J}}{\text{s}} \;/\; 180\,\text{MeV} \;/\; 1.602\times 10^{-13}\,\frac{\text{J}}{\text{MeV}} = 6.94\times 10^{10}\,\text{s}^{-1} \\ S &= R_{\text{fission}} \cdot 2.44 = 1.69\times 10^{11}\,\text{s}^{-1} \end{aligned} \tag{5.1}$$

5.3.1 *Verification of the Model*

To verify the present reactor model, simulations are carried out in which the spectrum in the reactor core and at different distances behind the exit of the radial channel is determined. Simulated spectra are available from the operating staff of the reactor, which have been validated experimentally [7], and are depicted in Fig. B.32 in the appendix. It should be noted that the plots represent absolute flux values.

In the course of this thesis, the simulations are repeated to verify the model and to determine the correct source strength for the normalization of further simulation results. For this purpose, the spectrum is determined on a circular area of radius $r = 1$ cm at the distances in question. This rather small radius is chosen because

[1] During the fission of ^{235}U, a total energy of 204 MeV is released whereby 180 MeV are ultimately converted into heat from kinetic energy of the fission fragments (167 MeV), the produced neutrons (5 MeV), and electrons of the beta-unstable fission products (8 MeV).

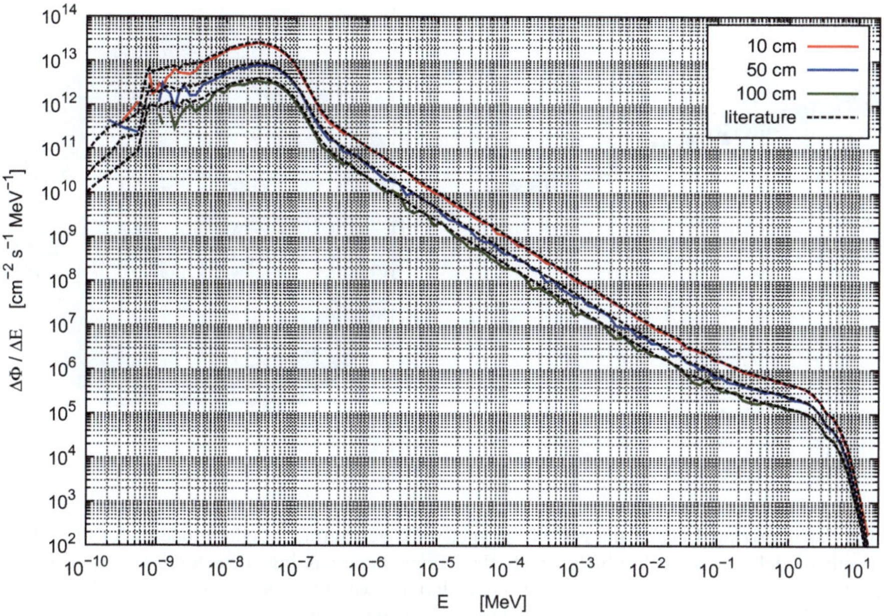

Fig. 5.3 Spectra at three positions behind the exit of the radial channel on its axis calculated with MCNP and literature values [7] from Fig. B.32. The absolute flux values in the cold energy region are rather low and exhibit a large relative error

the present spectra were measured on the channel axis and the flux distribution is assumed to be inhomogeneous over the entire cross-sectional area. The results of the simulation are shown as solid lines in Fig. 5.3. The literature values are drawn as dashed lines. The spectra were normalized to the bin-width so that a different binning can be chosen as for the literature values. The spectra are in good agreement and hence the MCNP model of the AKR-2 and the calculated source strength from Eq. 5.1 are considered valid.

5.4 Optimization of the Prototype Moderator

The prototype is optimized for use at the AKR-2 in order to carry out reasonable measurements even at the low neutron flux provided. For the simulation studies, the present MCNP model of the reactor is extended by the prototype moderator and the detector required for the measurements. In the following sections the simulation studies for optimizing the prototype design are outlined.

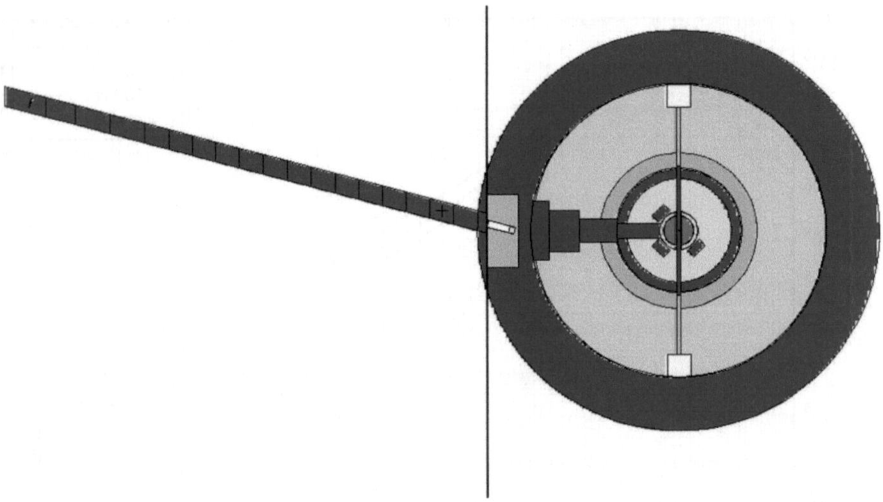

Fig. 5.4 Horizontal cut of the MCNP geometry for generating weight windows for the AKR-2 reactor model to increase the efficiency of calculations encompassing a detector in 4 m distance of the prototype moderator which is located 10 cm behind the exit of the radial experimental channel. The vertical black line represents an ideal absorber to shield the detector from neutrons not originating from the flux channel of the prototype

5.4.1 Generation of Weight Windows

As a result of the space conditions in the experimental hall of the AKR-2, the maximum distance of a detector from the exit of the radial channel amounts to 4 m. To optimize the design of the prototype, the neutron fluxes at this distance are to be determined. The divergence of neutrons leaving the radial channel and the prototype, respectively, leads to a significant reduction of the neutron flux at the detector. Since no evacuated neutron guides will be used for the measurements on the prototype, the neutrons are additionally scattered by air on the way to the remote detector so that the flux at the detector is reduced further. For this reason variance reduction techniques, such as Weight Windows, should be used to minimize the computational time of the parameter studies.

For this aim, the distance from the channel exit or the surface of the prototype, which is located 10 cm apart behind the channel exit, is divided into 18 segments with a length of 20 cm containing air. The simulation model is depicted in Fig. 5.4.

5.4.2 Optimization of the Thermal Moderator

Since the neutrons leaving the reactor core are at least partly thermalized, not the slowing-down power but a low absorption cross section is of particular importance

Table 5.1 Maximum thermal flux, its relative error, and its longitudinal position in the prototype moderator operating at the AKR-2 for beryllium, heavy water, and Graphite

Material	Φ_{th}^{max} [cm^{-2} s^{-1}]	σ_{rel} (%)	z [cm]
Beryllium	7.67×10^5	0.2	3.5
Heavy water	6.95×10^5	0.6	3.5
Graphite	6.19×10^5	0.9	2.5

for the moderator material of the prototype. Therefore, heavy water and graphite can be considered, whereby heavy water has a much higher moderating ratio according to Sect. 2.3.1, so that undermoderated neutrons from the core can still be thermalized while existing thermal neutrons are not absorbed. The use of hydrogen-based moderators would lead to higher absorption rate whereas beryllium is excluded due to its toxicity and high market price.

Nevertheless, simulations are carried out to determine the maximum thermal flux in a prototype made of beryllium, heavy water and graphite. The dimensions of the cylindrical moderator are chosen to be $h = 41.4$ cm and $r = 31$ cm. The results summarized in Table 5.1 show that Beryllium provides slightly better values than heavy water whereas the maximum flux in graphite is the lowest. The thermal flux maximum is found at a depth of 3.5 cm (Be, D_2O) and 2.5 cm (C), respectively. This shows that the neutrons from the reactor core are already strongly pre-moderated.

The geometry of the prototype moderator for the parameter study can be seen in Fig. 5.5. The radius of the heavy water tank is chosen to be $r = 31$ cm to cover the exit of the radial channel so that no neutrons are lost in the gap between the moderator and the channel exit due to their divergence. A channel is implemented into the heavy water tank, which is intended to serve as a slot for cold moderators as well as a thermal flux channel. The channel extends to the thermal flux maximum and has a radius of $r = 3$ cm. According to the studies shown in Sect. 4.2.5.2, a smaller flux channel has a larger thermal peak brilliance but it does not provide enough space for placing different cold moderators during the experiments. The flux channel is tilted by an angle of 15° with respect to the axis of the experimental channel of the reactor allowing the reduction of the fast neutron and gamma background originating from the reactor core.

For the investigation of the dependence of the flux at the detector position on the thickness of the moderator, two different approaches are followed. In the first pass, the angular dependence of the thermal flux for different moderator thicknesses is determined at the exit of the flux channel. The neutrons, which leave the flux channel at an angle of $\theta \leq 0.573°$ with respect to the axis of the flux channel, reach the detector with $r_{Detector} = 4$ cm at 4 m distance. An angle of $\theta \leq 0.678°$ corresponds to a detector position of 3.38 m, $\theta \leq 0.769°$ to 2.98 m, $\theta \leq 0.924°$ to 2.48 m and $\theta \leq 1.146°$ to 2 m.

This procedure, however, does not take into account the influence of the air in the path of the neutrons to the detector. For this purpose, a further study is carried out in

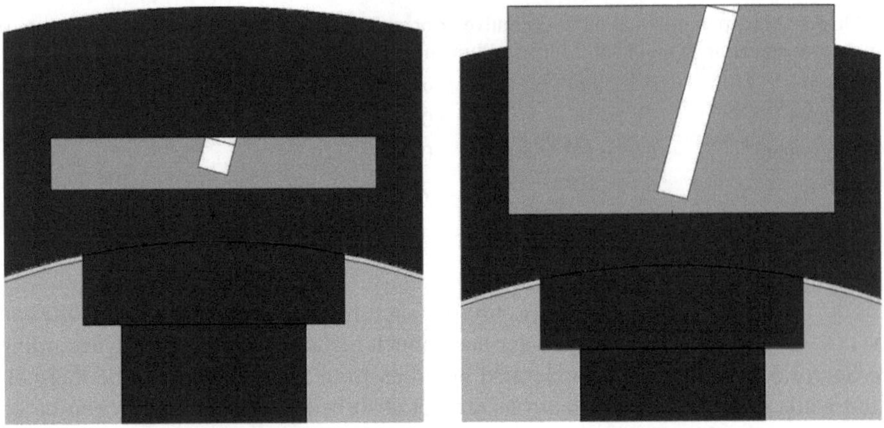

(a) Thickness 10 cm. (b) Thickness 40 cm.

Fig. 5.5 Excerpt of the simulation geometry for the determination of the optimum thickness of the cylindrical, D_2O-based prototype moderator with a thermal flux channel at the AKR-2

which a detector is defined in the MCNP model at 4 m distance, where the incoming thermal flux is determined in dependence to the moderator thickness by means of the simulation model depicted in Fig. 5.4.

According to Sect. 5.4.1, the path to the detector is divided into several air-filled segments so that weight windows can be used to increase the performance of the simulations. The surface of the moderator must be shielded in order to measure the effect of the flux channel increasing the peak brilliance which would otherwise be completely overlaid by the high background of thermal neutrons. Therefore, an ideal absorber is placed at the front face of the moderator, which provides a window for the flux channel.

The results of both approaches are shown in Fig. 5.6 in terms of the thermal neutron yield in dependence to the moderator thickness. Solid lines indicate the number of thermal neutrons at the flux channel exit for different distances of the detector, represented by different angles with respect to the flux channel axis. Due to the fact that the neutrons from the reactor core are already strongly pre-moderated, a considerable thermal flux is achieved even with low moderator thicknesses. As the thickness increases, the thermal flux increases in the forward direction at the finger exit up to a thickness of 35 cm, from which the thermal flux does not increase further. The shape of the different curves are identical but the values are higher for larger detector distances or opening angles, respectively, which is governed by the inverse-square law. The results of the second approach are shown by the dashed lines. The results of both approaches agree very well, apart from statistical fluctuations in small and large moderator thicknesses, if the air in the model is removed. A comparison of the two dashed curves shows, however, that the influence of air is not negligible. When using air, the flux at the detector position decreases by 28% averaged over all moderator dimensions.

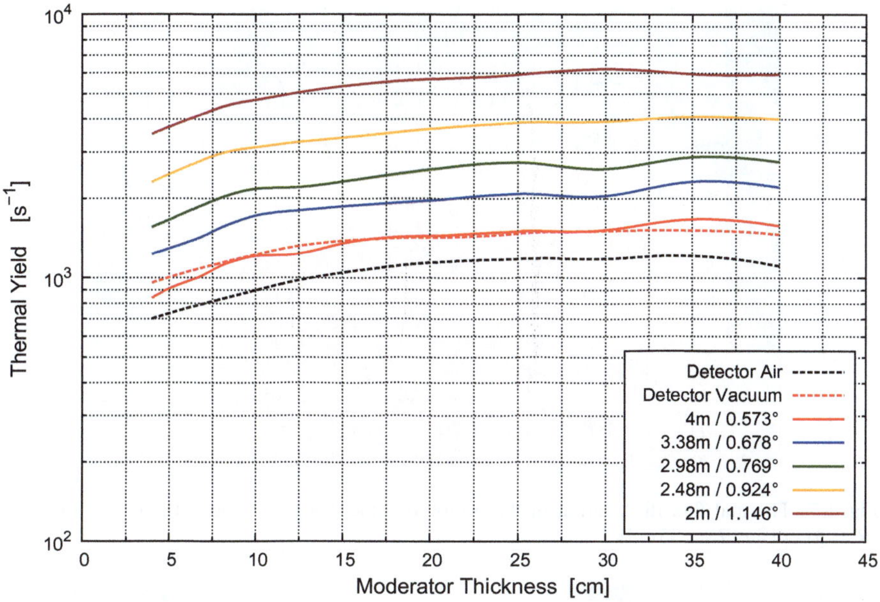

Fig. 5.6 Thermal yield at the detector ($r = 4\,\text{cm}$) in 4 m distance in air and vacuum (dashed lines) and number of thermal neutrons leaving the flux channel with different maximum emission angles with respect to the channel axis, which corresponds to different detector distances (solid lines). The relative error of most values is below 5%

Based on the results presented herein, the cylindrical tank of the prototype moderator will have a radius of $R = 31\,\text{cm}$ and a length of $L = 35\,\text{cm}$. The spatial distribution of the thermal flux in this configuration is depicted in Fig. 5.7.

When carrying out the experiments, the moderator surface should be completely shielded with the exception of the flux channel exit as discussed above. In additional experiments the brilliance should also be measured on the surface of the bulk moderator in order to determine the increase in peak brilliance by using flux channels. For this purpose the flux channel should be filled with heavy water or the shielding should be rotated so that the exit of the shielding no longer fits the flux channel, but the bare surface of the heavy water tank.

5.4.3 *Optimization of the Cold Moderator*

The dimensions of the cold moderator are optimized for the use of parahydrogen. For this aim, a cylindrical volume filled with para-H_2 is defined at the bottom of the flux channel, see Fig. 5.8. Its dimensions r and h are varied independently of each other. The flux of cold neutrons within the energy range $0.38\,\text{meV} < E < 13.1\,\text{meV}$ at the detector is determined for each configuration.

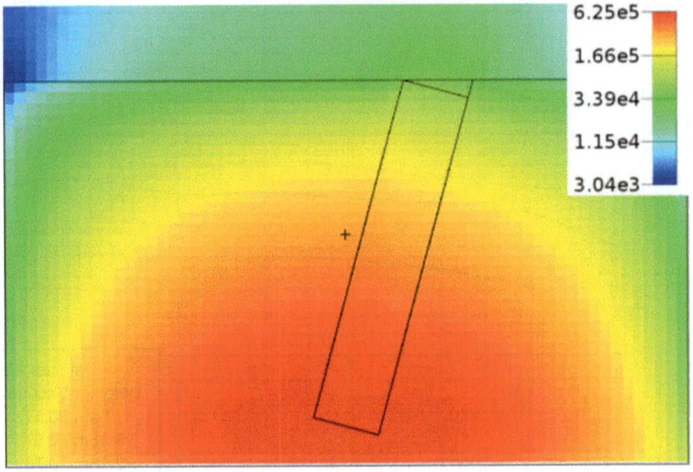

Fig. 5.7 Thermal flux distribution in the prototype moderator with the optimum configuration ($R = 31$ cm and $L = 35$ cm)

(a) Cold moderator dimensions $r = 2$ cm and $l = 16$ cm in a channel with $r = 3$ cm.

(b) Cold moderator dimensions $r = 4$ cm and $l = 10$ cm in a channel with $r = 4$ cm.

Fig. 5.8 Excerpt of the simulation geometry for the determination of the optimum dimensions of the cylindrical, parahydrogen-based cold moderator in the flux channel at the AKR-2

Figure 5.9 shows the total flux of cold neutrons reaching the detector in 4 m distance in dependence to the length of the cold moderator for various radii. It is evident that the optimum length is in the range between 10 cm and 15 cm, depending on the radius. These values are in accordance to the conclusions in Sect. 2.3.2 regarding the mean free path of cold neutrons in parahydrogen. Consequently, the length of the

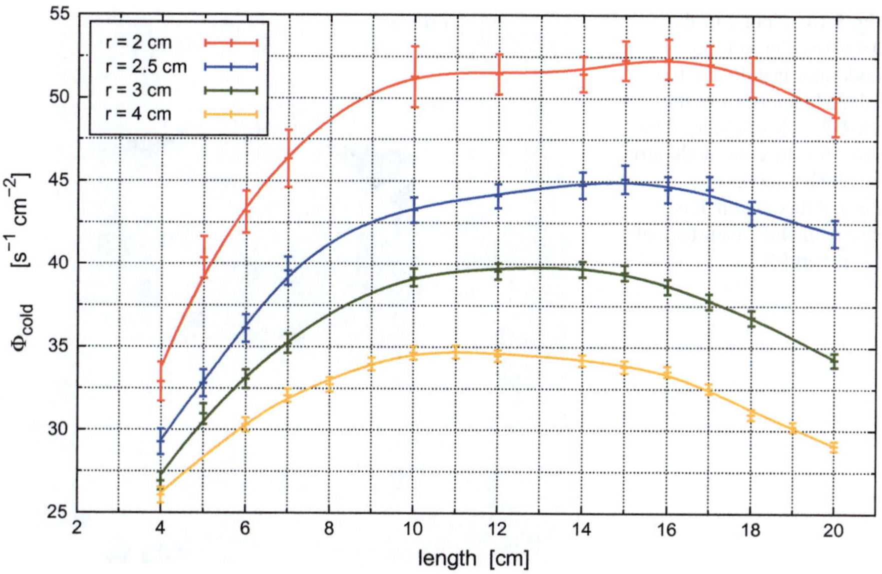

Fig. 5.9 Cold flux at the detector ($r = 4$ cm) in 4 m distance in air normalized to the cold moderator outer surface in dependence to the cold moderator length for different radii according to Fig. 5.8. The relative error of most values remains below 3%

cold moderator will be $l = 10$ cm whereas the radius is set to $r = 3$ cm. In this case the detector is supplied with 1107 cold neutrons per second sufficient to measure a precise spectrum of cold neutrons even when using choppers with 10–20% duty cycle.

5.5 Experiments

The experiments are carried out and evaluated by [8]. The final setup of the prototype moderator and the measuring technique as well as the preliminary results are explained in the following sections.

5.5.1 *Final Design of the Thermal Moderator*

The heavy water is kept in an aluminum tank due to the good processability, favorable material properties and neutron economy of aluminum. The alloy specification reads AlMg3Al. Based on the simulation results explained in Sect. 5.4.2 aiming at the optimization of the prototype moderator, the drum is chosen to have an inner radius of $r = 31$ cm and an inner height of $h = 35$ cm. The lateral thickness of the drum

Fig. 5.10 Picture of the D_2O-based prototype moderator installed at the radial channel exit of the AKR-2 with shielding. The insert for the cold moderator, which also serves as a thermal flux channel, is clearly visible. Courtesy of T. Cronert [8]

amounts to 3 mm whereas the thickness of lid and bottom is 10 mm. The drum is equipped with an ISO-K 100 flange and an ISO-K 63 flange (Ref. DIN 28404, ISO 1609) to accommodate the cryogenic moderators. The flanges also serve as thermal flux channels if the cold moderators are removed.

The heavy water is recycled cooling water originating from Canadian CANDU reactors. Its purity is stated to be >98% which could be confirmed by measurements at the ZEA-3 of Forschungszentrum Jülich. Figure 5.10 shows a photograph of the D_2O drum with PE shielding in its mounting at the exit of the radial channel of the AKR-2. The two flanges extend through the shielding.

5.5.2 Design of the Cold Moderator

In addition to the cold moderator also the cooling system must be designed for the measurements at the AKR-2 [3]. Innovative approaches are required compared to common cryogenic engineering in order to be able to precisely adjust and also to stabilize different hydrogen compositions at low temperatures.

Adjustable Ortho-/Parahydrogen Cryo System An adjustable ortho- to parahydrogen cryo system is designed in order to be able to adjust the neutron spectrum to the demands of the experiment. According to Sect. 2.3.2 the mixture of orthohydrogen with parallel proton spins forming three higher energy states and parahydrogen with antiparallel proton spins will tend to its thermodynamic equilibrium ratio, which is determined by the temperature. At ambient temperatures the ortho-to-parahydrogen

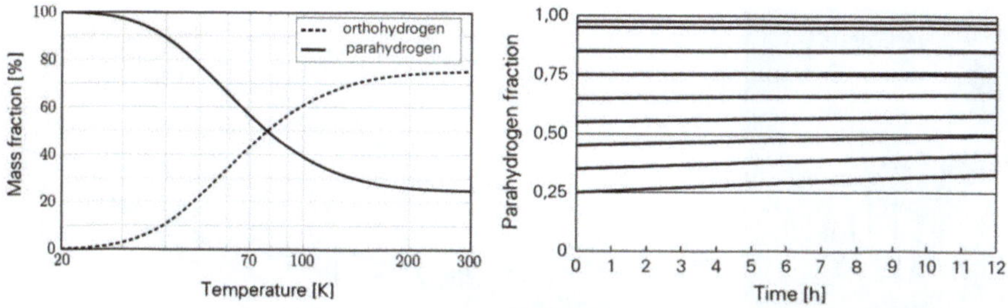

(a) Thermodynamic equilibrium of ortho- and para-hydrogen mixtures in dependence to the temperature.

(b) Time dependence of the parahydrogen fraction due to auto conversion in liquid hydrogen at 17 K and 150 kPa.

Fig. 5.11 Properties of the ortho- and para fraction in hydrogen. Courtesy of M. Klaus [3]

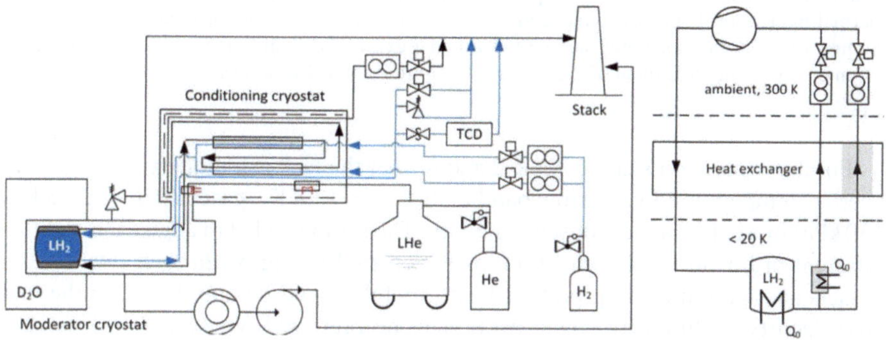

(a) Proof of principle.

(b) Closed hydrogen system for an HBS facility.

Fig. 5.12 Setups of the cryogenic system with adjustable ortho- to parahydrogen composition. The moderator vessel (approx. 0.31 liquid H_2) will be cooled externally by a helium cold gas to 17 K. Two tube-in-tube heat exchangers cool down incoming hydrogen with the exhausting helium before it finally condenses in the moderator vessel. One of the two heat exchangers will be filled with catalyst material, converting to almost pure parahydrogen. Courtesy of T. ronert [8]

ratio amounts to 3:1 whereas below 20 K the equilibrium is shifted to a parahydrogen fraction of nearly one as indicated by Fig. 5.11a.

When the temperature changes, an autocatalytic conversion occurs caused by dipole-dipole-interactions of two orthohydrogen molecules changing the direction of the thermodynamic equilibrium [9] which is a rather slow decelerating process. For higher orthohydrogen contents than equilibrium the autocatalysis has to be antagonized, cf. Fig. 5.11b, which is achieved by feeding hydrogen of a higher orthohydrogen concentration to the system. The conversion can be significantly increased by using paramagnetic substances like transition metals and their oxides [10] used in liquefiers as catalysts.

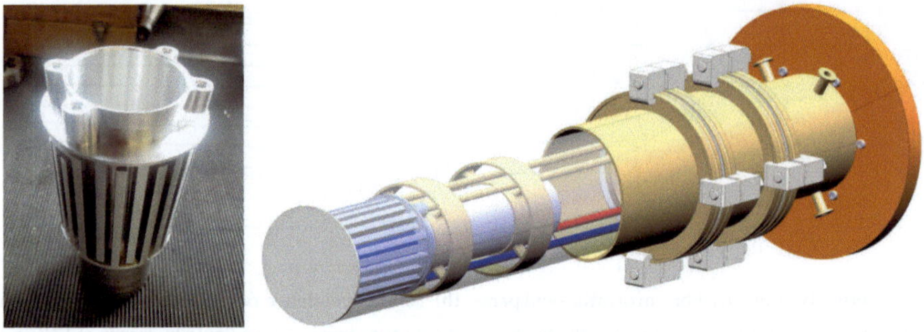

Fig. 5.13 Cold source prototype as developed and built for tests at the AKR-2. The cold source is a cylindrical aluminum vessel ($r = 25\,\text{mm}$, $l = 100\,\text{mm}$) to condense H_2 or CH_4 gas, surrounded by a He gas cooling labyrinth. The working pressure is defined as 150 kPa for liquid H_2. The cold source is installed on a standard flange and can be inserted into an evacuated beam tube in the thermal moderator. Four gas connections are used for input and output of moderator gas and He cooling gas. On the moderator surface towards the exit of the beam tube, a filter crystal can be mounted for further treatment of the spectrum. Courtesy of T. Cronert [8]

For the experiments at the AKR-2 and further the task is to liquefy hydrogen and keep its temperature in a narrow band (solid: $T < 13.8\,\text{K}$, gaseous: $T > 21.8\,\text{K}$ at 150 kPa) in a safe manner for several hours. The setup needed to adjust and stabilize different ortho- to parahydrogen ratios is depicted in Fig. 5.12a with two vacuum vessels and helium cooling provided by a mobile liquid helium dewar. In the final HBS facility a closed hydrogen cycle may be preferred whose possible setup is depicted in Fig. 5.12b.

Thermally and neutron induced para-to-ortho conversion becomes increasingly important for higher neutron fluxes and heat loads. These mechanics, however, are notoriously difficult to simulate, so that experimental results from the prototype can be extrapolated to larger sources (such as the ESS).

Prototype Cold Moderator At the HBS each instrument can be provided with an individual cold source designed to the specific needs of the corresponding instrument. The shape of the cold source can be adapted to the beam size and divergence desired at the sample position. The temperature and material can be chosen to fulfill the spectral demands of the experiment as described in Sect. 2.4.

As depicted in Fig. 5.13 the cold moderator is placed inside the thermal flux channel, so that the cold source is directly placed within the thermal flux maximum. The dimensions of the cold source are designed according to the principle of low-dimensional beam extraction proposed for the ESS cold moderator [11, 12] and the mean free path of cold and thermal neutrons in 100% parahydrogen. The design is optimized for liquid parahydrogen and ortho-parahydrogen mixtures, but the condensation of gaseous methane and liquid mesitylene is also possible.

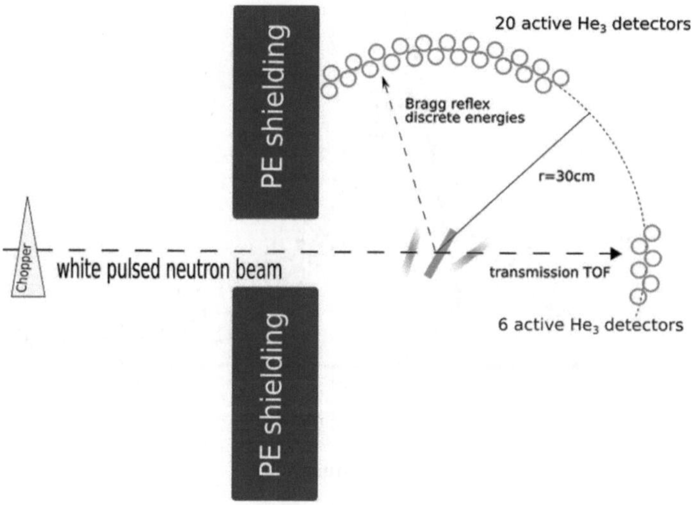

Fig. 5.14 Instrumentation for measurements on the prototype moderator at the AKR-2 reactor. Courtesy of T. Cronert [8]

5.5.3 *Measurement Setup*

The instrumentation for most measuring series for both the thermal flux channel and the cold moderator is shown schematically in Fig. 5.14. A chopper is placed directly behind the channel exit to allow the flight time measurement of the neutrons despite the continuous operation of the AKR-2. Within 2–3 m distance to the chopper, depending on the specific measurement, a monochromator crystal is placed on a goniometer.[2] According to the Bragg condition, neutrons of specific discrete energies are reflected, depending on the angle between the lattice planes of the monochromator and the direction of incidence of the neutrons as elucidated in Sect. 2.4. To detect the reflected neutrons, twenty He_3 Geiger tubes are used, which are positioned on a circular sector at a distance of 30 cm perpendicular to the beam axis.

Neutrons whose energy or wavelength, respectively, does not satisfy the Bragg condition are transmitted almost unimpeded. The energy of these neutrons is determined by means of a TOF measurement with six additional He_3 Geiger tubes located on the beam axis. As indicated in Sect. 5.4.2, the flight path between chopper and monochromator is shielded by borated PE to minimize the neutron background from the moderator surface and the reactor itself.

[2] A goniometer is used to either measure an angle or allow an object to be rotated to a precise angular position.

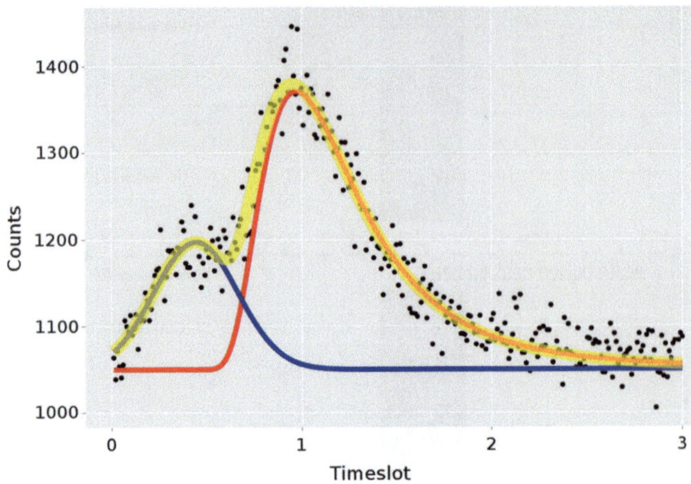

Fig. 5.15 Results of the TOF measurements at the thermal flux channel of the prototype moderator at the AKR-2. The blue curve represents the chopper function which corresponds to the fast neutrons. The red curve is a fitted Maxwell-Boltzmann distribution of the thermal neutrons. The maximum of the Maxwell-Boltzmann distribution located at time slot 1 corresponds to an energy of 55 meV whereas its mean energy amounts to 25 meV. Courtesy of T. Cronert [8]

5.5.4 Preliminary Results

This section presents the preliminary results of the experiments conducted so far. In this context, the TOF measurements of the transmitted neutrons in the case of a thermal channel and the corresponding measurement with the use of a cold mesitylene moderator are discussed. The remaining measurement results are still under evaluation and will be discussed in [8] after publication.

Thermal Flux Channel The spectrum of transmitted neutrons determined by TOF measurements at the thermal flux channel are depicted in Fig. 5.15. The absolute counts of this eight-hour measurement are depicted as a function of the flight time of the neutrons. Thermal neutrons are fitted by a Maxwell-Boltzmann distribution, whose maximum occurs in time slot 1, corresponding to an energy of 55 meV. The mean of the Maxwell-Boltzmann distribution amounts to 25 meV corresponding to room temperature. A second, smaller maximum is visible in earlier time slots, which is caused by the fast neutrons detected shortly after the chopper opens. This maximum is fitted as a Gaussian distribution which is called the chopper function.

Since in this representation the chopper function and the distribution of the raw values have not yet been deconvoluted, each value is subject to an uncertainty which corresponds to the FWHM of the chopper function. In addition, the energy-dependent detector efficiency of the He_3 Geiger tubes has not yet been taken into account. Therefore, a much more precise statement about the neutron spectrum can be made after final evaluation of the data.

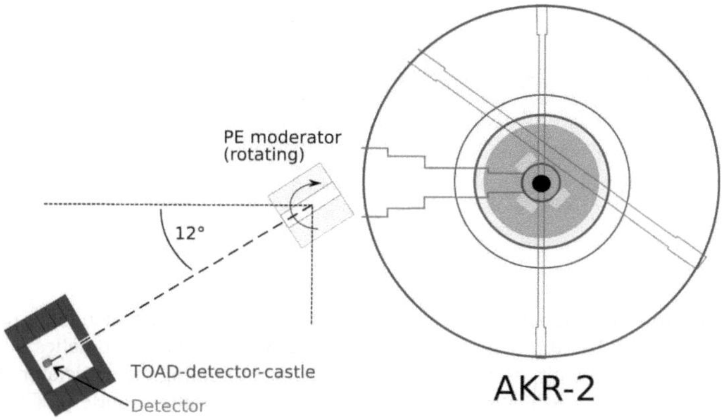

Fig. 5.16 Experimental setup for measuring the angular dependence of the thermal brilliance of a flux channel in the PE based prototype moderator at the AKR-2. The detector setup is depicted in detail in Fig. 5.14. Courtesy of T. Cronert [8]

In addition to the spectrum of the neutrons leaving the moderator through the flux channel, the directional dependence of the brilliance with respect to the finger axis is also investigated in further experiments. It is not possible to freely position the detector setup (cf. Sect. 5.5.3) due to a lack of space. Instead of the detector, the moderator is rotated in the horizontal plane in order to determine the directional dependence of the thermal flux channel. Furthermore, due to the size and the high mass of the prototype moderator, no precise orientation can be set, which would lead to a large systematic error with respect to the angle. Instead, a cylindrical moderator consisting of PE with a diameter of 40 cm and a height of 40 cm is used. A bore with a diameter of 9 cm reaches through the center of the PE cylinder and serves as a thermal flux channel (Fig. 5.16). For the positioning of the PE cylinder no goniometer is used because of the high mass. Instead a laser sight and a meter scale is used. The systematic uncertainty of the adjusted angles is estimated to $\pm 1\%$.

With this experimental setup, the flux of thermal neutrons is measured depending on the angular orientation of the moderator, i.e., the flux channel. The results are depicted in Fig. 5.17. In conclusion the experimental results meet the expectations obtained by the comprehensive simulation work. At an angle for which the detector lies on the axis of the flux channel, most neutrons are detected. The edges around this maximum are slightly asymmetric, which can be explained by the relatively high absorption cross section of PE: For small angles, the neutrons leaving the radial channel of the AKR-2 must effectively travel a shorter distance in the PE before reaching the detector. It can be concluded that flux channels can significantly increase the yield of thermal neutrons in a specific directions caused by the streaming effect. In the future more experiments will be conducted which will be focused on the study of the angular dependence of the brilliance.

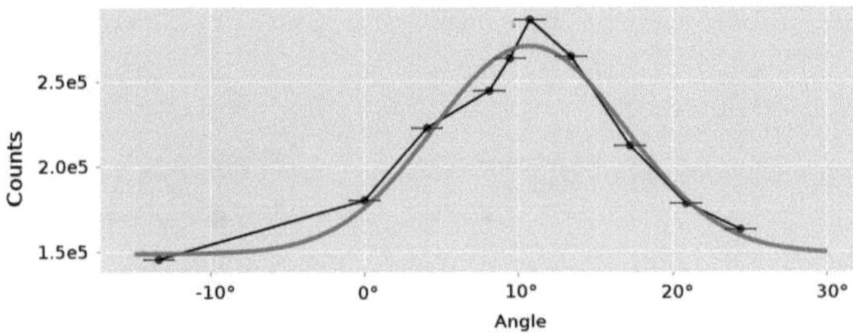

Fig. 5.17 Results of the measurements of the angular dependence of the thermal brilliance of the flux channel in the PE based prototype moderator at the AKR-2. The uncertainties of the angular values arise from the difficulty of precisely positioning and orienting the moderator. The smooth curve denotes a Gaussian distribution fitted to the experimental results. Courtesy of T. Cronert [8]

Table 5.2 Simulated flux values and measured count rates of a fission chamber at different positions in the thermal flux channel of the prototype moderator at the AKR-2. The fission chamber is sensitive below 0.456 eV, therefore only neutrons with corresponding energies are taken into account in the simulation

Position	Φ^{sim} [cm^{-2} s^{-1}]	σ_{rel} (%)	$\Phi^{sim}/\Phi^{sim}_{0\,cm}$	cts^{lit}/100 s	cts^{lit}/cts$^{lit}_{0\,cm}$
0 cm	6.74×10^5	0.4	1.000	38317	1.000
10 cm	5.01×10^5	0.4	0.743	27210	0.710
20 cm	2.74×10^5	0.6	0.407	14766	0.385
30 cm	8.52×10^4	1.2	0.126	4685	0.122
40 cm	2.25×10^4	2.5	0.037	1266	0.033

Cold Moderators Measurements on the hydrogen-based cold moderator aiming for a verification of the simulation results discussed in Sects. 3.3.3 and 4.3 will be carried out with the ortho-para mixing cryostat in the near future. The spectra for the mesitylene-based moderator have a very bad statistical quality. However, it can be stated that the cold neutron flux with cryogenic mesitylene (20 K) is twice as high as with thermal mesitylene (300 K). This result is consistent with the experiences from [13] confirming the validity of the models.

Fission Chamber Measurements in the Thermal Channel To verify the results of the measurements so far and the AKR simulation model as well, an absolute measurement of the thermal flux in the thermal channel of the prototype moderator is performed at the department. For this purpose, a fission chamber[3] is used, which is located at different positions inside and outside the thermal channel on its axis. The fission chamber is sensitive to neutrons with energies below 0.456 eV, which corresponds to a temperature of 10584 K. In order to be independent of the detector efficiency of the fission chamber, the results are normalized to the value from the first measurement position. The results are shown in Table 5.2.

[3]Ionization chambers whose electrode is coated with enriched uranium are suitable for detecting slow neutrons and are called fission chambers.

The position 0 cm corresponds to the fission chamber being at flux channel bottom. The simulated values correspond to the flux in the volume occupied by the fission chamber while the actual measured values represent the count rate per 100 s measuring time. A very good agreement between the simulation and the absolute measurement is evident when comparing the normalized values. The results of the simulations aiming for optimizing the thermal moderator discussed in Sect. 5.4.2 are therefore considered as valid.

References

1. Institute of Power Engineering Training Reactor. Training reactor AKR-2. Description of the reactor facility, procedure of operation. Technical Report (Technical University Dresden, 2005). https://tu-dresden.de/ing/maschinenwesen/iet/wket/ressourcen/dateien/akr2/Lehrmaterialien/aufbaue.pdf. Accessed 18 Oct 2016
2. Institute of Power Engineering Training Reactor, TU Dresden. Ausbildungskernreaktor. Lehre—Forschung—Information. Booklet. https://tu-dresden.de/ing/maschinenwesen/iet/wket/ressourcen/dateien/akr2/Lehrmaterialien/flyer.pdf. Accessed 19 Oct 2016
3. U. Rücker et al., The Jülich high-brilliance neutron source project. Eur. Phys. J. Plus **131**(1), 19 (2016). https://doi.org/10.1140/epjp/i2016-16019-5
4. J.P. Dabruck et al., Development of a moderator system for the high brilliance neutron source project, in *IlNuovo Cimento C 38.192*, 8 May 2016. https://doi.org/10.1393/ncc/i2015-15192-0. https://www.sif.it/riviste/ncc/econtents/2015/038/06/article/18
5. C. Lange, J. Gonzales, MCNP-Eingabefile AKR. MCNP Input file of the AKR-2 Reactor. TU Dresden, Chair for Hydrogen and Nuclear Energy Technology, 16 May 2012
6. X-5 Monte Carlo Team. MCNP—A General N-Particle Transport Code, Version 5. LA-UR-03-1987. Technical report Los Alamos National Laboratory, 1 Feb 2008. https://laws.lanl.gov/vhosts/mcnp.lanl.gov/pdf_files/la-ur-03-1987.pdf
7. W. Hansen, R. Schneider, Erweiterung und Verifikation von TRAMO zur Lösung von gekoppelten Neutron/Gamma-transportproblemen und Überprüfung von Kerndatenbibliotheken. Final Report DFG Project. Project Number HA-3009/1-1. Technical University Dresden, 27 Feb 2002
8. T. Cronert, High brilliance thermal and cold neutron moderators–an experimental validation of MCNP-simulations for the HBS-project. Ph.D. thesis. Jülich: Forschungszentrum Jülich, Jülich Centre for Neutron Science 2
9. Y. Ya. Milenko, R.M. Sibileva, M.A. Strzhemechny, Natural ortho-para conversion rate in liquid and gaseous hydrogen. J. Low Temp. Phys. **107**(1), 77–92 (1997). ISSN: 1573-7357. https://doi.org/10.1007/BF02396837
10. D.H. Weitzel, W.V. Loebenstein, J.W. Draper, O.E. Park, Ortho-para catalysis in liquid-hydrogen production. J. Res. Natl. Bur. Stand. **60**(3) (1958). https://doi.org/10.6028/jres.060.026. http://nvlpubs.nist.gov/nistpubs/jres/60/jresv60n3p221_A1b.pdf
11. K. Batkov, A. Takibayev, L. Zanini, F. Mezei, Unperturbed moderator brightness in pulsed neutron sources, in *Nuclear Instruments Methods A: Accelerators, Spectrometers, Detectors and Associated Equipment*, vol. 729, pp. 500–505 (2013). ISSN: 0168-9002. https://doi.org/10.1016/j.nima.2013.07.031. http://www.sciencedirect.com/science/article/pii/S0168900213010073
12. F. Mezei et al., Low dimensional neutron moderators for enhanced source brightness. J. Neutron Res. **172**, 101–105 (2014). https://doi.org/10.3233/JNR-140013. https://arxiv.org/pdf/1311.2474
13. V. Ananiev et al., Pelletized cold moderator of the IBR-2 reactor: current status and future development, in *Journal of Physics: Conference Series*, vol. 746, No. 1 (2016). https://doi.org/10.1088/1742-6596/746/1/012031

Chapter 6
Conclusions

6.1 Summary

Comprehensive simulation studies on the target and the thermal and cold moderator-reflector assembly aiming for an optimization of the neutronic performance of laser- or compact accelerator-driven neutron sources are conducted in the scope of this thesis. The well-established Monte Carlo particle transport code MCNP is employed for the simulations.

Laser-based ion acceleration has made enormous progress in the recent years and will become a serious alternative to conventional accelerators. For such systems a moderator system is developed based on the experimental findings at the Trident Laser in Los Alamos, USA. For this purpose, the energy, angular, and time distributions of the accelerated deuterons and the generated neutrons, respectively, are implemented as a source definition in MCNP. Due to the relatively large divergence of the generated neutrons, it is not feasible to use compact hydrogen-based moderators such as polyethylene or light water. Instead, beryllium is proposed and its design and dimensions are optimized in extensive parameter studies to maximize the thermal neutron flux. Cryogenic moderators consisting of parahydrogen, solid methane, or mesitylene are embedded in the thermal moderator and optimized regarding the arrangement in the thermal moderator and their dimensions. During these simulations, a new approach is followed in which not only the cold moderators are positioned in the center of the thermal moderator, but also the thermal neutrons are extracted from the thermal flux maximum of the thermal moderator with the aid of flux channels. This configuration is referred to as the Finger Moderator. For neutron scattering experiments it is relevant how many neutrons can ultimately be extracted and transported to the sample, which is measured in terms of the brilliance. For some types of experiments, e.g., time of flight techniques, the temporal structure of the neutron pulses is also important. The effects of an additional graphite reflector on these performance parameters (peak brilliance and pulse structure) are investigated. It is shown that by using the aforementioned flux channels, the peak brilliance can be significantly increased compared to earlier designs.

J. P. Dabruck, *Target Station Optimization for the High-Brilliance Neutron Source HBS*, Springer Theses, https://doi.org/10.1007/978-3-030-05639-1_6

Although laser-based systems can be built considerably more compact, existing facilities are still subject to development. Thus, it is expected that a stable operation providing neutrons for scientific users is hard to achieve in the medium term. Thus, it is decided to rely on a conventional particle accelerator. For this, the neutron yield is determined as a function of the incident particle type and energy (protons and deuterons between 5 and 100 MeV) as well as the target material (beryllium and lithium). Further simulations are conducted assuming a proton accelerator with an energy of 50 MeV and beryllium as the target material. With 50 MeV protons reasonable neutron yields are achieved at foreseeable costs for accelerator and shielding. The availability of nuclear data for the simulations with protons is very good. However, the simulation results can also be scaled to deuteron operation as soon as the corresponding nuclear databases for deuterons are verified. Under these conditions, the target is designed with a maximum surface power density of $1\,\mathrm{kW/cm^2}$. The shape is chosen in such a way that the target is as compact as possible for this given surface power density. A backing plate made of vanadium prevents hydrogen embrittlement of the active target material.

For technical reasons, simulations involving other particles besides neutrons, electrons, and photons can only be inadequately parallelized. Therefore, a new method is developed for the optimization studies on the moderator system which uses the MCNP function `PTRAC`. The phase space coordinates (location, direction, energy, and time) of neutrons produced in the target are recorded and stored in an external file which is read line by line in subsequent simulations by a specifically developed MCNP subroutine and uses the stored data as source particles. This method can, in principle, also be applied to other issues, such as simulations of nuclear reactors with heterogeneous reactor cores. Instead of simulating fission processes in the fuel, the neutrons can be started, e.g., at the inner surface of the reactor pressure vessel and the model can be simplified considerably, which improves the simulation performance.

The optimization studies for the moderator system are carried out efficiently with the previously generated source file. Since the source of the primary neutrons is more compact and less divergent than in laser-driven systems, a compact polyethylene moderator is used, which is surrounded by a low-absorption reflector made of beryllium. The shape and dimensions of the moderator and reflector are optimized to maximize the flux of thermal neutrons. Thermal flux channels and two cold moderators made of parahydrogen and solid methane, respectively, are implemented into the moderator-reflector assembly and the peak brilliance and the pulse structure as a function of the reflector thickness are determined at each channel in order to be able to select the optimum dimensions for each type of experiment.

In order to confirm the simulation results, especially the concept of the thermal flux channels, experiments are carried out at the AKR-2 training reactor at TU Dresden, Germany. The prototype moderator system consists of a heavy water-based premoderator with a slot for various cold moderators. To increase the effectiveness of the measurements, the moderator system is optimized in various parameter studies. For this purpose, an MCNP model of the reactor is provided by the operator of the AKR-2. In the course of this thesis, the reactor model is verified using experimental data. The D_2O moderator and the embedded cold moderator are designed to achieve the maximum thermal and cold flux at the detector, respectively. Weight Windows are created and applied as a variance reduction technique to accelerate the simulations. The experimental data evaluated so far show a good agreement with the simulations and thus confirm the beneficial effects of the flux channels.

6.2 Outlook

In this thesis the moderator system of a medium-flux CANS is optimized with regard to the neutronic performance. The suggested design can of course not exactly be transferred due to the technical and mechanical constraints dictated by the material properties and the operational parameters. Thus, the presented results shall serve as a basis for a system operating under realistic conditions and shall demonstrate which physical aspects have to be considered to maximize the neutron yield available to the instruments. After the target station has been conceptualized according to the technical and mechanical aspects, a comprehensive simulation has to be carried out, which encompasses the final structures in all details in order to assess the actual performance of the entire system.

In laser-driven systems, the current bottleneck is the development of a continuously operating target for laser-based ion acceleration. A further technical challenge is to integrate the converter, in which the neutrons are produced, into the moderator in order to minimize neutron loss in backward direction. If compact converters can be placed inside the moderator, it is recommended to use a moderator design similar to the proposed accelerator-driven system, for example, a compact thermal polyethylene moderator with a low-absorption reflector.

For the first accelerator-based HBS the work to be done is much more specific. The accelerator will be selected from a precise cost calculation for the accelerator itself and the shielding taking into account the achievable neutron yield. The space requirements of the site must also be considered. A change in the accelerator parameters also requires a detailed redesign of the target-moderator-reflector assembly. In this regard, lead or steel as a reflector materials should also be investigated. Even if the moderation performance is not as favorable as for beryllium, they shield the gamma radiation generated in the target, which reduces the need for additional shielding measures. The purchase costs of steel is also significantly lower than that of beryllium. Regarding the thermal moderator, not only the classical materials but also hydrogen-doped metals or graphite are conceivable. The concentration of dopant represents a further parameter in the optimization of the moderator performance. An example is $ZrH_{1.65}$, for which $S(\alpha, \beta)$ nuclear data is available. For other concentrations of dopant and materials, e.g., BeH_x, CH_x, ZrD_x, experimental measurements are absolutely essential. For the novel Flux Finger approach it should be investigated if the brilliance can be further increased by burnishing the inner surfaces of the flux channels.

Depending on the accelerator performance, the use of a stationary target may not be indicated due to an excessive surface power density. However, a rotating target can drastically reduce the surface power density, but would also affect the moderator geometry due to its size. Anyway, a full re-design is necessary if the final concept of the target is specified. Additionally, detailed studies on the DPA-production have to be carried out besides the assessment of the effects of thermal stress and hydrogen embrittlement for each target configuration.

For cold moderators it has been shown that further studies on aromatic substances are needed. Due to the poor availability of nuclear data, experiments are inevitable. Furthermore, it has been found that pure parahydrogen moderators lead to undermoderation of the neutrons due to the characteristic behavior of the cross section, as a result of which the neutrons do not reach thermal equilibrium with the moderator. Mixtures of para- and orthohydrogen may improve this situation and are the subject of current investigations.

In order to optimally adjust the neutron source to all important types of scattering experiments, the operation of several target stations which are supplied by a single accelerator and a beam multiplexer may be advantageous. Each target station could be operated with specially adapted repetition rates, e.g., 25–50 Hz, 100 Hz, and 400 Hz and corresponding pulse lengths and duty cycles, respectively. This also clearly results in different optimum dimensions of the associated moderator-reflector setup. Thereby, the implementation of fast and epithermal beamlines into the assembly should be considered to serve additional types of experiments like neutron imaging or activation analysis. Also conceivable are experiments whose optical axes are not horizontal so that additional flux channels above or below the horizontal channels may be implemented.

To allow for a precise assessment of the performance of the individual instruments, simulations with specialized codes such as McStas [1] or Vitess [2] are to be carried out which use as input parameters the phase space coordinates of the neutrons entering the beamlines. This information is supplied as single-event data, similar to the `PTRAC` output, in the *MCPL* (Monte Carlo Particle List) file format [3]. Since these codes are not capable of determining the background consisting of fast neutrons and gammas at the sample and detector locations, separate MCNP calculations are required which involve the geometry of the neutron source and its shielding as well as the setup of the instrumentation.

With regard to the experiments on the AKR-2, further simulations should be carried out, which represent the complete measurement setup in detail and higher resolution. Based thereon, definitive conclusions can be drawn about the quality of the nuclear data of the employed cryogenic moderator materials.

References

1. K. Lefmann, K. Nielsen, McStas, a general software package for neutron ray-tracing simulations. Neutron News 10(3), 20–23 (1999). https://doi.org/10.1080/10448639908233684
2. G. Zsigmond et al., A survey of simulations of complex neutronic systems by VITESS. Nucl. Instrum. Methods Phys. Res. Sect. A Accel. Spectrometers Detect. Assoc. Equip. **529**(1), in *Proceedings of the Joint Meeting of the International Conference on Neutron Optics (NOP2004) and the Third International Workshop on Position-Sensitive Neutron Detectors (PSND2004)* (2004), pp. 218–222. ISSN: 0168-9002. https://doi.org/10.1016/j.nima.2004.04.205. http://www.sciencedirect.com/science/article/pii/S0168900204009015
3. T. Kittelmann et al., Monte Carlo Particle Lists: MCPL. Comput. Phys. Commun. **218**, 17–42 (2017). ISSN: 0010-4655. https://doi.org/10.1016/j.cpc.2017.04.012. http://www.sciencedirect.com/science/article/pii/S0010465517301261

Appendix A
Data and Cross Sections

A.1 Cross Sections

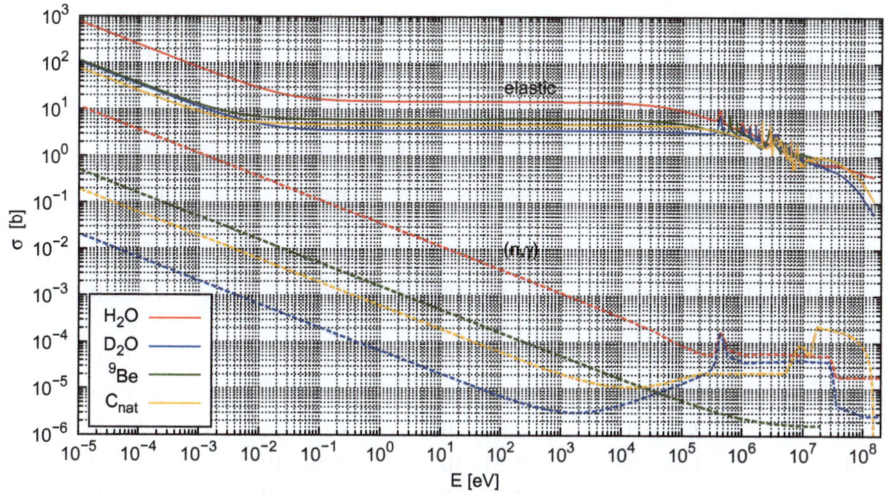

Fig. A.1 Elastic and radiative capture cross sections of typical moderator materials in dependence to the incident neutron energy evaluated at a moderator temperature of 293.15 K. Data is taken from the ENDF/B-VII.1 database [9]

J. P. Dabruck, *Target Station Optimization for the High-Brilliance Neutron Source HBS*, Springer Theses, https://doi.org/10.1007/978-3-030-05639-1

Table A.1 Neutron energies ranging from ultra-cold to fast neutrons and corresponding values for velocity, wavelength, and temperature

E (MeV)	E (meV)	v (m/s)	λ (Å)	T (K)
1.00×10^{-13}	1.00×10^{4}	4.4	904	0.0012
2.69×10^{-12}	0.0027	23	174	0.031
1.40×10^{-11}	0.014	52	77	0.16
7.25×10^{-11}	0.072	118	34	0.84
2.00×10^{-10}	0.20	196	20	2.32
3.76×10^{-10}	0.38	268	15	4.36
5.68×10^{-10}	0.57	330	12	6.59
1.72×10^{-9}	1.72	574	6.89	20
2.61×10^{-9}	2.61	707	5.60	30
3.45×10^{-9}	3.45	812	4.87	40
4.55×10^{-9}	4.55	933	4.24	53
5.22×10^{-9}	5.22	1000	3.96	61
6.89×10^{-9}	6.89	1148	3.44	80
1.01×10^{-8}	10.12	1392	2.84	117
1.31×10^{-8}	13.08	1582	2.50	152
2.05×10^{-8}	20.50	1980	2.00	238
2.35×10^{-8}	23.52	2121	1.86	273
2.53×10^{-8}	25.25	2198	1.80	293
3.00×10^{-8}	30.00	2396	1.65	348
5.25×10^{-8}	52.53	3170	1.25	610
1.27×10^{-7}	127.00	4929	0.80	1474
2.73×10^{-7}	272.56	7221	0.55	3163
1.41×10^{-6}	1414.20	1.65×10^{4}	0.24	1.64×10^{4}
3.81×10^{-5}	3.81×10^{4}	8.54×10^{4}	0.046	4.42×10^{5}
1.98×10^{-4}	1.98×10^{5}	1.94×10^{5}	0.020	2.29×10^{6}
0.001	1.03×10^{6}	4.43×10^{5}	0.0089	1.19×10^{7}
0.005	5.32×10^{6}	1.01×10^{6}	0.0039	6.17×10^{7}
0.028	2.76×10^{7}	2.30×10^{6}	0.0017	3.20×10^{8}
0.14	1.43×10^{8}	5.23×10^{6}	7.56×10^{-4}	1.66×10^{9}
3.85	3.86×10^{9}	2.72×10^{7}	1.46×10^{-4}	4.47×10^{10}
20.00	2.00×10^{10}	6.19×10^{7}	6.40×10^{-5}	2.32×10^{11}

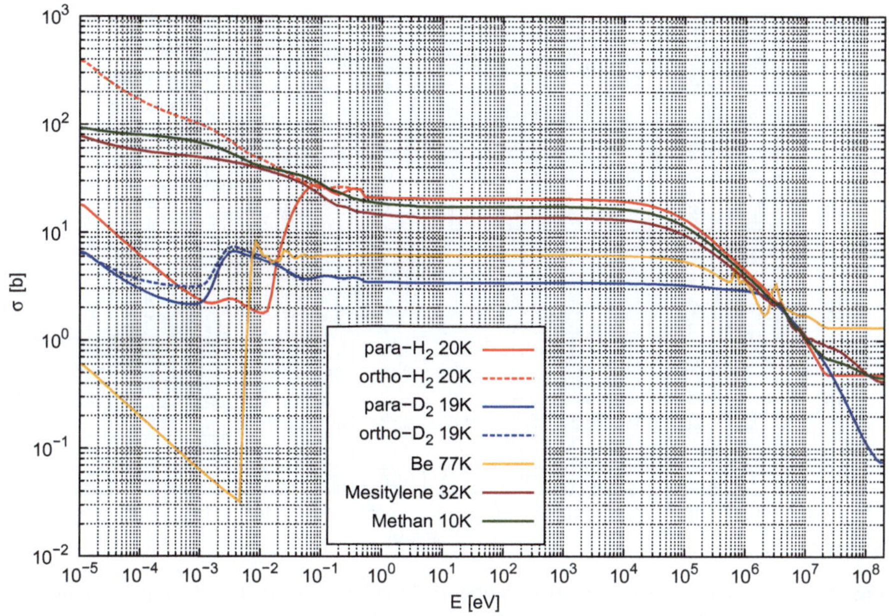

Fig. A.2 Scattering cross sections of typical cold moderator materials in dependence to the incident neutron energy evaluated at different temperatures. Data derived from the ENDF/B-VII database except for mesitylene which is taken from [1] and methane taken from [2]

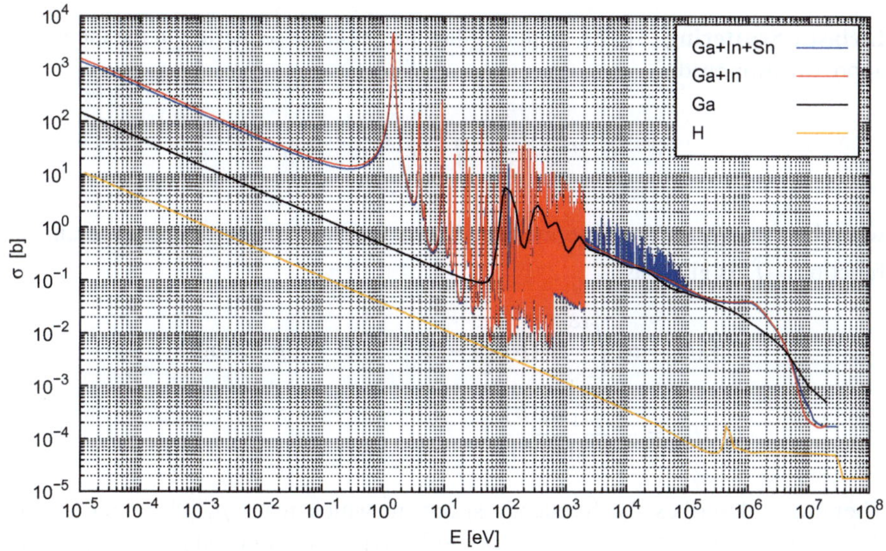

Fig. A.3 (n,γ) radiative capture cross section of gallium and two common gallium eutectics taken from ENDF/B-VII.1 [9]

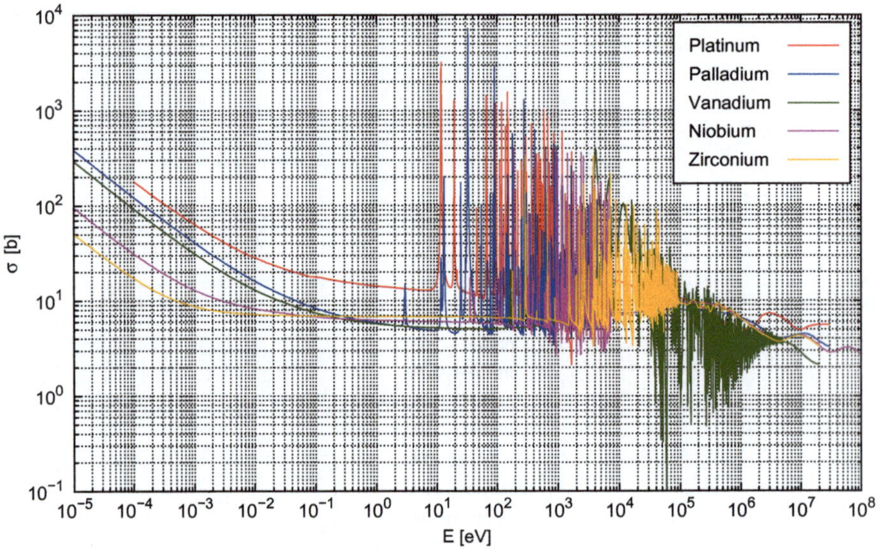

Fig. A.4 (n,γ) radiative capture cross section of different materials for the target backing. The natural isotope composition for each material is considered

A.2 Formulas

Inelastic Scattering [3] The scattering law for inelastic scattering in the incoherent approximation reads

$$S^{\mathrm{inel}}(\alpha,\beta) = \frac{\mathrm{e}^{-\beta}}{2\,\pi} \int_{-\infty}^{\infty} \mathrm{e}^{\mathrm{i}\,\beta\,\hat{t}}\,\mathrm{e}^{-\gamma(\hat{t})}\,\mathrm{d}\hat{t} \quad , \tag{A.1}$$

whereby $\hat{t} = \frac{\hbar}{k_{\mathrm{B}}\,T}$ is the time measured as a function of the temperature and the function $\gamma(\hat{t})$ is given as

$$\gamma(\hat{t}) = \alpha \int_{-\infty}^{\infty} P(\beta)\left[1 - \mathrm{e}^{-\mathrm{i}\,\beta\,\hat{t}}\right]\mathrm{e}^{-\beta/2}\,\mathrm{d}\beta \quad \text{, with}$$
$$P(\beta) = \frac{\sum_{j=1}^{K}\rho_j(\beta)}{2\,\beta\,\sinh(\beta/2)} \quad . \tag{A.2}$$

Thereby, ρ describes the frequency spectrum which reads $\rho_j(\beta) = w_j\,\delta(\beta_j)$ for discrete oscillators, $\rho_j(\beta) = \rho_{\mathrm{s}}(\beta)$ for a solid-type spectrum and $\rho_j(\beta) = \rho_{\mathrm{t}}(\beta)$ for a translational spectrum. The translational spectrum may either contain a free-gas law or a diffusion spectrum occurring in liquids.

Phonon Expansion [3] Taking into account the solid-type spectrum ρ_s for the scattering law in Eq. A.1, the exponential can be expanded as

$$e^{-\gamma_s(\hat{t})} = e^{-\alpha\,\lambda_s} \sum_{n=0}^{\infty} \frac{1}{n!} \left[\alpha \int_{-\infty}^{\infty} P_s(\beta) e^{-\beta/2}\, e^{i\,\beta\,\hat{t}}\, d\beta \right] \tag{A.3}$$

where λ_s is called the Debey-Waller coefficient which is a function of $P_s(\beta)$ from Eq. A.2 encompassing the lattice effects in crystalline media.

Incoherent Elastic Scattering [3] For hydrogenous solids the zero-th component of the phonon series in Eq. A.3 corresponds to scattering without energy transfer and interference and is thus referred to as incoherent elastic scattering for which the scattering function reads

$$S_{\mathrm{inc}}^{\mathrm{el}}(\alpha, \beta) = e^{-\alpha\,\lambda}\, \delta(\beta) \quad . \tag{A.4}$$

Coherent Elastic Scattering [3] For crystalline solids the zero-th phonon mode describes coherent elastic scattering without energy transfer and interference. The corresponding cross section is given by

$$\sigma_{\mathrm{coh}}(E, \mu) = \frac{\sigma_{\mathrm{free}}}{E} \sum_{E_i < E} f_i\, e^{-4\,W\,E_i}\, \delta(\mu - \mu_i) \quad \text{, with} \tag{A.5}$$

$$\mu_i = 1 - \frac{E_i}{E} \quad ,$$

where the Bragg edges $E_i = \frac{\hbar^2\,\tau_i^2}{8\,m}$ are a function of the length τ_i of the corresponding vector of the reciprocal lattice of the solid and f_i represent crystallographic structure factors.

Liquid Hydrogen and Deuterium [3] As the spins of para- and orthohydrogen and -deuterium are not arbitrarily distributed they have to be treated differently from Eq. A.1. The explicit quantum-mechanical scattering law

$$S_{\mathrm{p/o}}(\alpha, \beta) = \sum_{J}^{\infty} P_J \times \frac{4\,\pi}{\sigma_{\mathrm{free}}} \left[A_{\mathrm{p/o}} \sum_{J'=0,2,4,\ldots} + B_{\mathrm{p/o}} \sum_{J'=1,3,5,\ldots} \right] (2\,J' + 1) \tag{A.6}$$

$$\times\; S_{\mathrm{f}}(w\,\alpha, \beta + \beta_{J\,J'}) \tag{A.7}$$

$$\times \sum_{l=|J'-J|}^{J'+J} 4\, J_l^2(y)\, C^2(J\,J'\,l; 00) \tag{A.8}$$

was considered by [4]. For parahydrogen and paradeuterium the summation is carried out over even values of J whereas for orthohydrogen and orthodeuterium over odd values. The parameters A and B are functions of the coherent and incoherent scattering lengths, whose shape depends on the substance in question being ortho- or para- hydrogen or deuterium. A statistical weight factor is represented by P_j and $\beta_{JJ'}$ is the energy transfer for a rotational transition, $j_l(x)$ a spherical Bessel function, $C(JJ'l; 00)$ the Clebsch-Gordan coefficients, and y is a function of the interatomic distances and molecule mass in the substance.

Appendix B
Additional Simulation Results

B.1 Laser-Driven Neutron Source

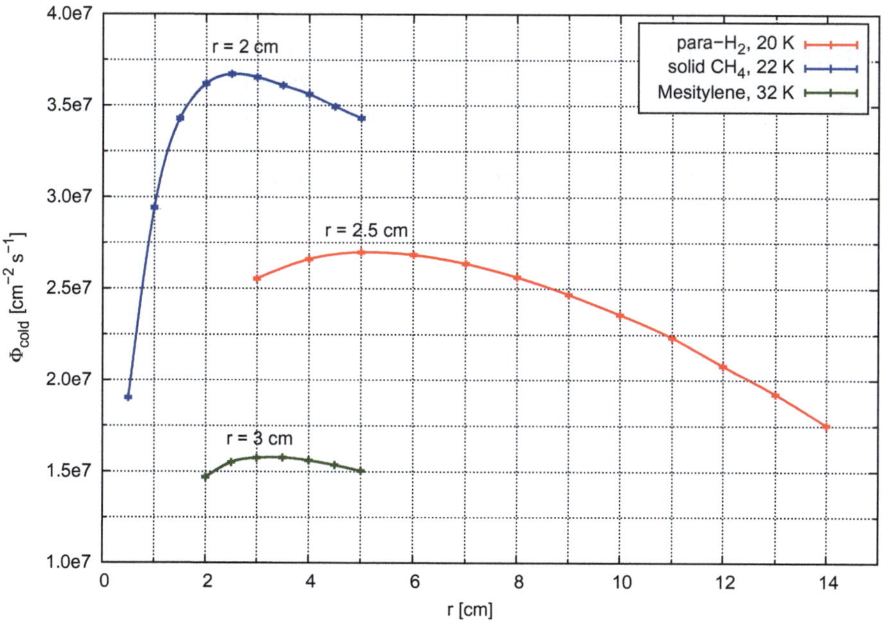

Fig. B.1 Flux of cold neutrons with $E < 7\,meV \mathrel{\widehat{=}} 81\ K$ on the outwardly directed cold moderator surface depending on its length. The optimum radius according to Table B.1 is taken into account. Values are given in terms of the neutron production at Trident operating at 10 Hz

J. P. Dabruck, *Target Station Optimization for the High-Brilliance Neutron Source HBS*, Springer Theses, https://doi.org/10.1007/978-3-030-05639-1

Table B.1 Flux of cold neutrons with $E < 7$ meV $\widehat{=}$ 81 K on the outwardly directed cold moderator surface depending on the radius for parahydrogen, solid methane and mesitylene. For each radius the optimum length is listed. Values are given in terms of the neutron production at Trident operating at 10 Hz

Moderator	r (cm)	l_{opt} (cm)	Φ_{cold} (s^{-1} cm^{-2})	$\sigma_{rel}(\Phi_{cold})(\%)$
para-H_2, 20 K	1.0	4	2.027×10^7	0.55
	1.5	5	2.415×10^7	0.34
	2.0	5	2.632×10^7	0.24
	2.5	5	2.696×10^7	0.19
	3.0	5	2.678×10^7	0.16
	3.5	6	2.617×10^7	0.14
	4.0	6	2.522×10^7	0.12
	4.5	6	2.411×10^7	0.11
	5.0	6	2.303×10^7	0.10
solid CH_4, 22 K	1.0	2.0	2.864×10^7	0.46
	1.5	2.5	3.477×10^7	0.27
	2.0	2.5	3.671×10^7	0.20
	2.5	2.5	3.667×10^7	0.16
	3.0	2.5	3.534×10^7	0.13
	3.5	2.5	3.346×10^7	0.12
	4.0	2.5	3.143×10^7	0.11
	4.5	2.5	2.936×10^7	0.10
	5.0	2.5	2.728×10^7	0.09
Mesitylene, 32 K	1.0	3.0	1.024×10^7	0.77
	2.0	3.0	1.525×10^7	0.31
	3.0	3.5	1.576×10^7	0.20
	4.0	3.0	1.461×10^7	0.15
	5.0	3.0	1.295×10^7	0.13

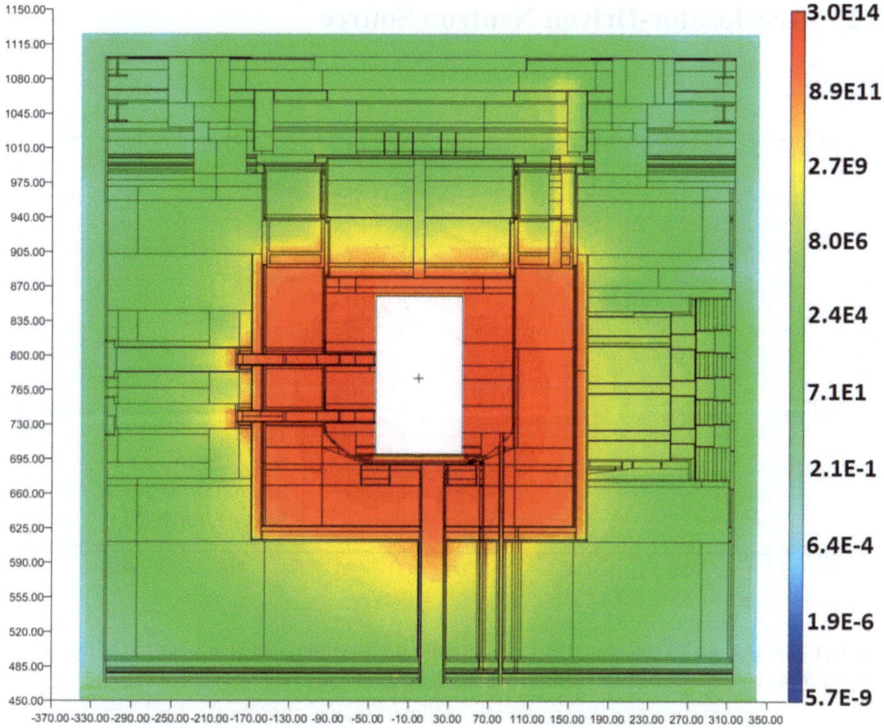

Fig. B.2 Neutron flux distribution in $cm^{-2}\,s^{-1}$ of the German FRJ-2 research reactor. The streaming effect can be seen very well at the pipes which break through the biological shield

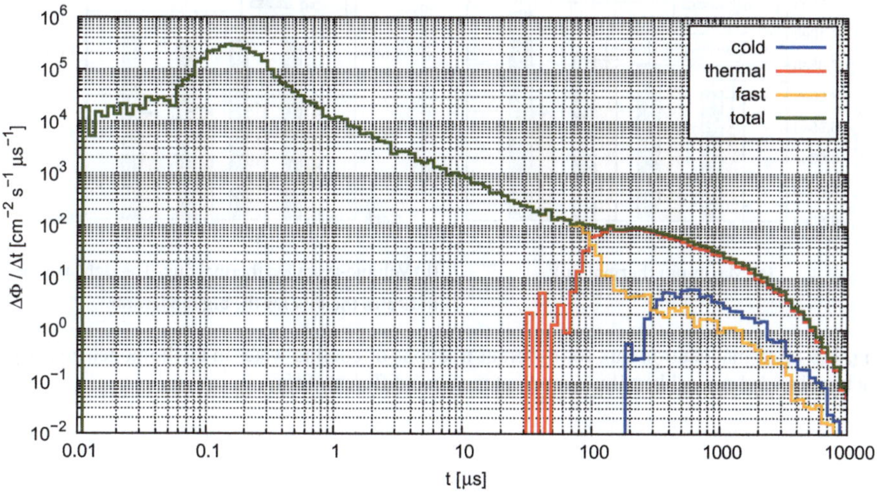

Fig. B.3 Pulse shape of cold, thermal and fast neutrons at the exit of the channel $\varphi = 90°$ in a 1° cone in forward direction. Values are normalized to Trident operating at 10Hz

B.2 Accelerator-Driven Neutron Source

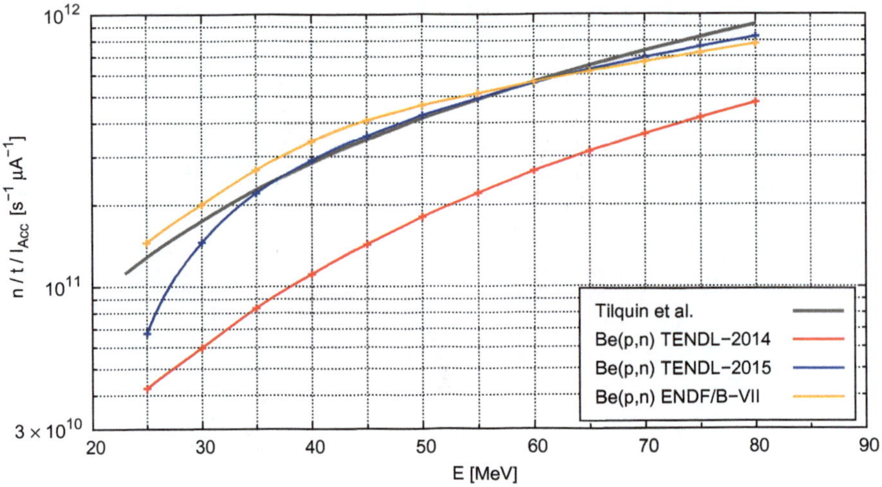

Fig. B.4 Neutron yields calculated by [5] of protons on beryllium in dependence to the ion energy. An empirical curve derived from measurements of Tilquin et. al is also included [6]

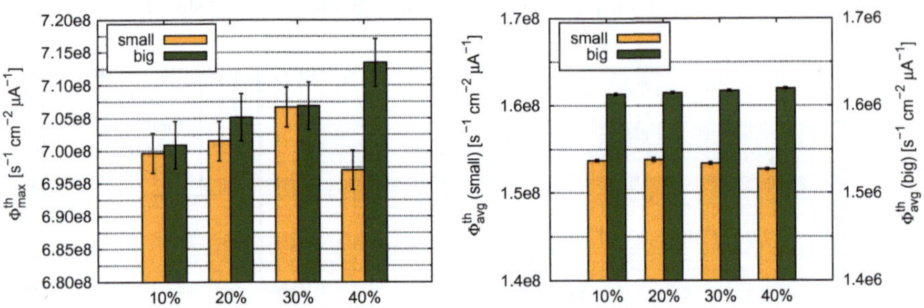

(a) Maximum thermal flux.

(b) Volume-averaged thermal flux. The right axis label indicates the flux for the big moderator.

Fig. B.5 Thermal neutron flux for different fractions of cooling water in the homogenized target backing in a cylindrical beryllium moderator of two different sizes

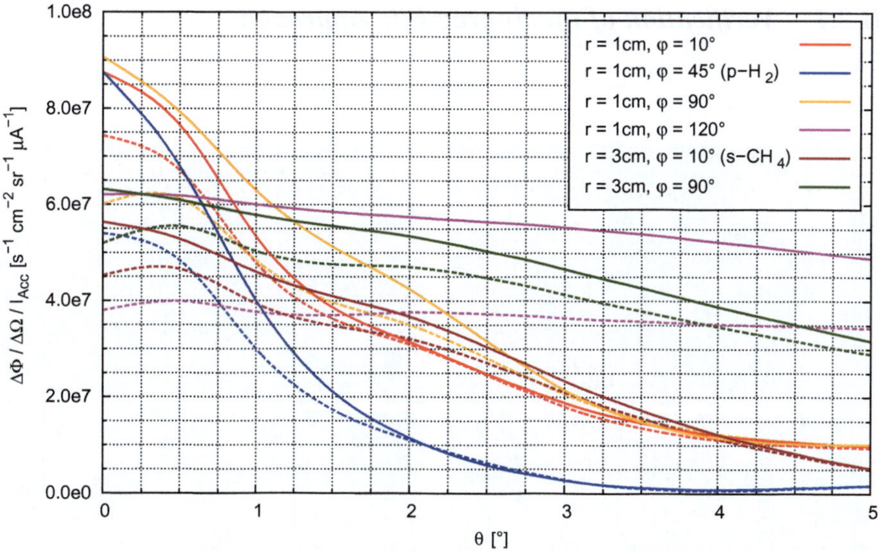

Fig. B.6 Thermal or cold brilliance, respectively, at the channel exits of the model with original channel positions (solid lines) and for a configuration in which the channels originate at the surface of the PE moderator (dashed lines)

B.2.1 Verification of the Source File Approach

(a) Deviation for 10^5 particles.

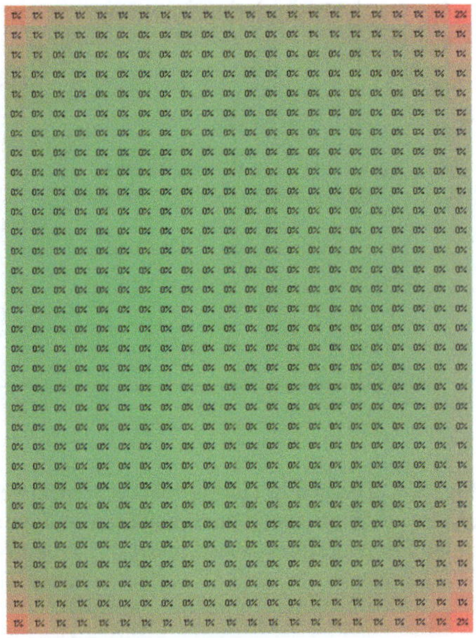

(b) Relative error for 10^5 particles.

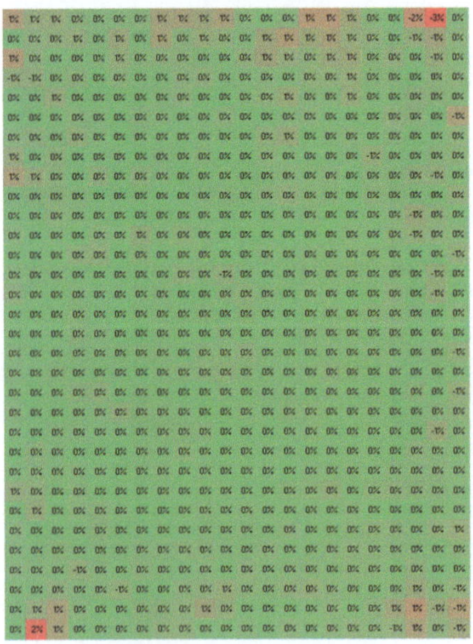

(c) Deviation for 10^6 particles.

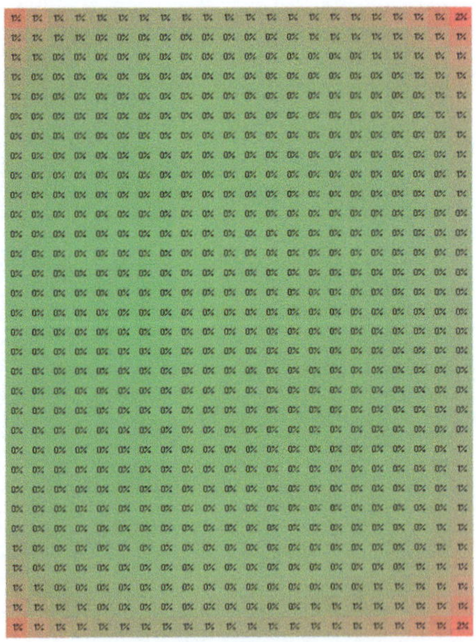

(d) Relative error for 10^6 particles.

Fig. B.7 Deviation of the spatial neutron flux distributions calculated using source files of different sizes and a full simulation with deuterons as starting particles and its relative error

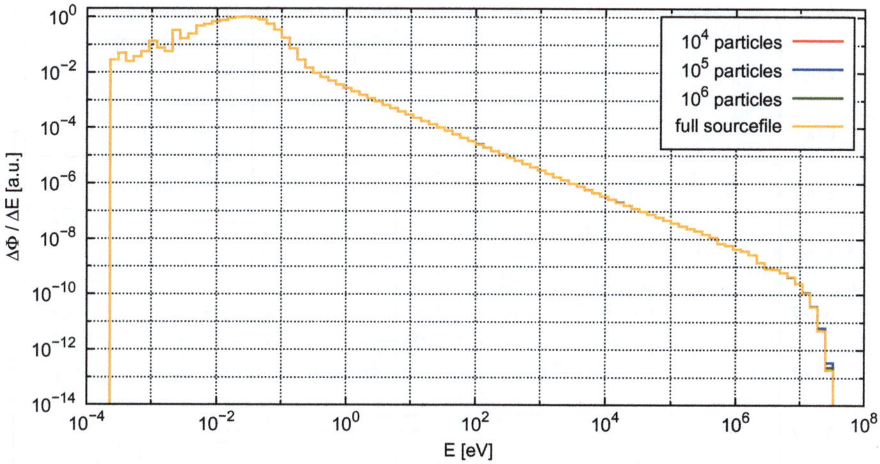

Fig. B.8 Spectra of neutrons normalized to their maximum in the beryllium cylinder from source files with different sizes. The full source file encompasses 2.6×10^6 particle entries

B.2.2 Extraction of Thermal Neutrons

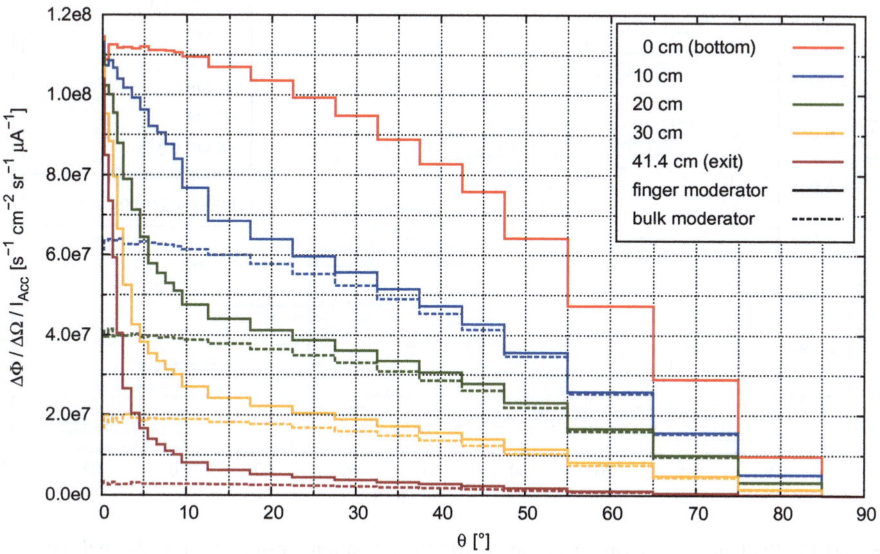

Fig. B.9 Brilliance of thermal neutrons at different positions inside the flux channel with $r = 1$ cm, $\varphi = 45°$ (solid lines) and a bulk moderator, for which the channel is filled with material up to the measuring position (dashed lines)

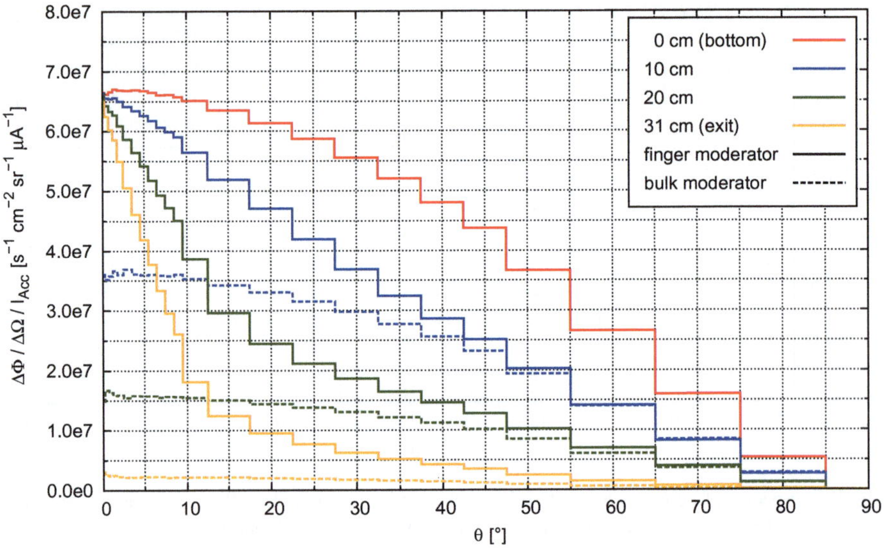

Fig. B.10 Brilliance of thermal neutrons at different positions inside the flux channel with $r = 1\,\text{cm}$, $\varphi = 90°$ (solid lines) and a bulk moderator, for which the channel is filled with material up to the measuring position (dashed lines)

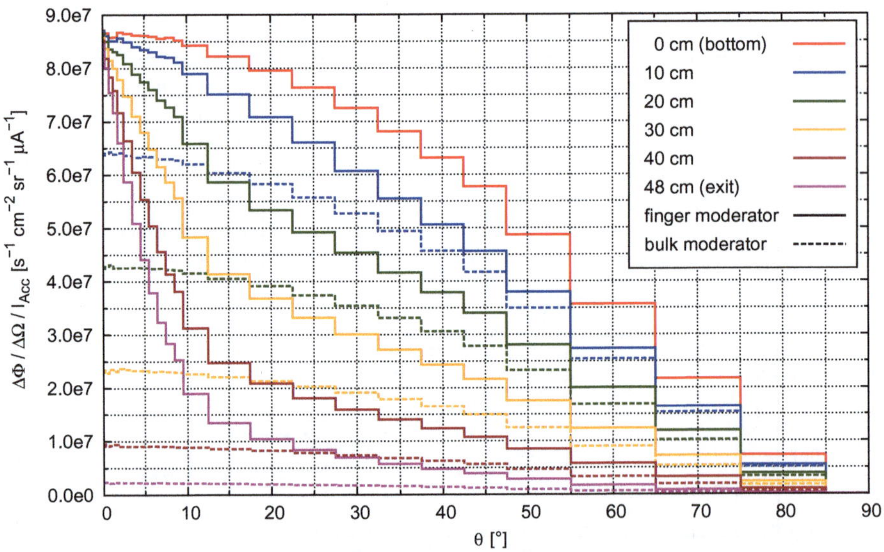

Fig. B.11 Brilliance of thermal neutrons at different positions inside the flux channel with $r = 3\,\text{cm}$, $\varphi = 10°$ (solid lines) and a bulk moderator, for which the channel is filled with material up to the measuring position (dashed lines)

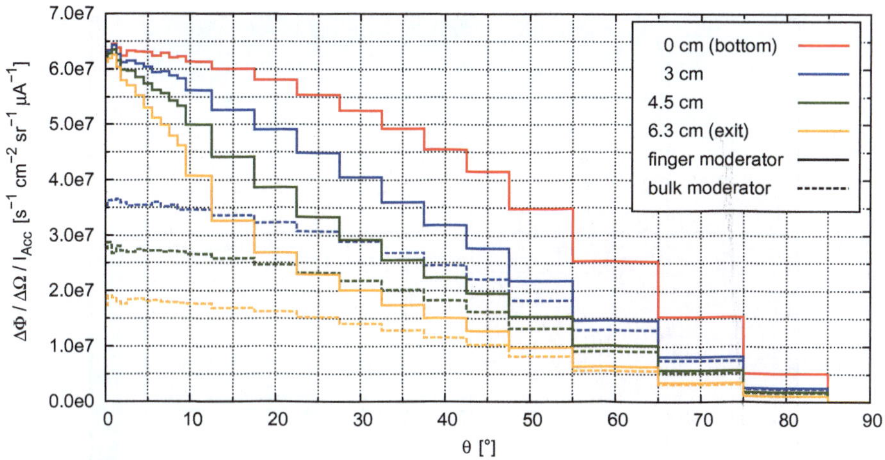

Fig. B.12 Brilliance of thermal neutrons at different positions inside the flux channel with $r = 1\,\text{cm}$, $\varphi = 120°$ (solid lines) and a bulk moderator, for which the channel is filled with material up to the measuring position (dashed lines)

B.2.3 Energy-Resolved Brilliance

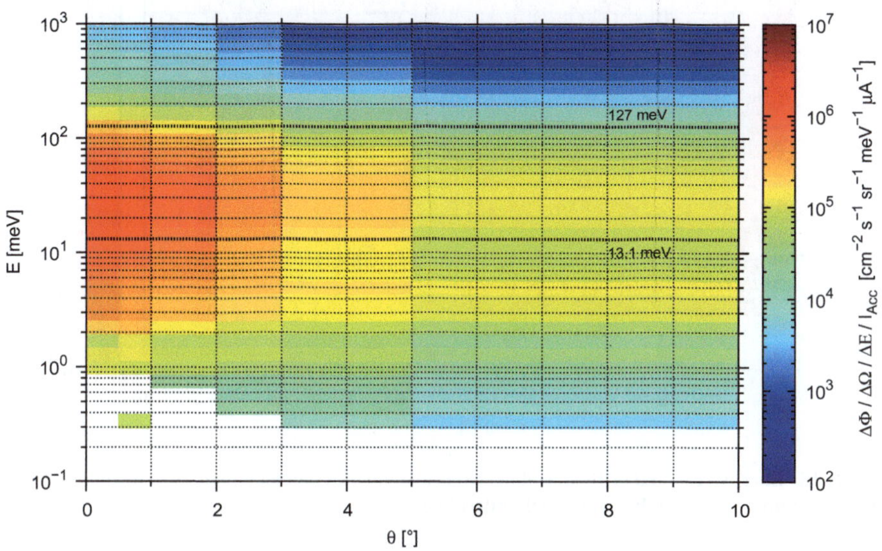

Fig. B.13 Energy-resolved brilliance of neutrons leaving the channel with $r = 1\,\text{cm}$, $\varphi = 90°$. The thermal regime is marked by horizontal dotted lines

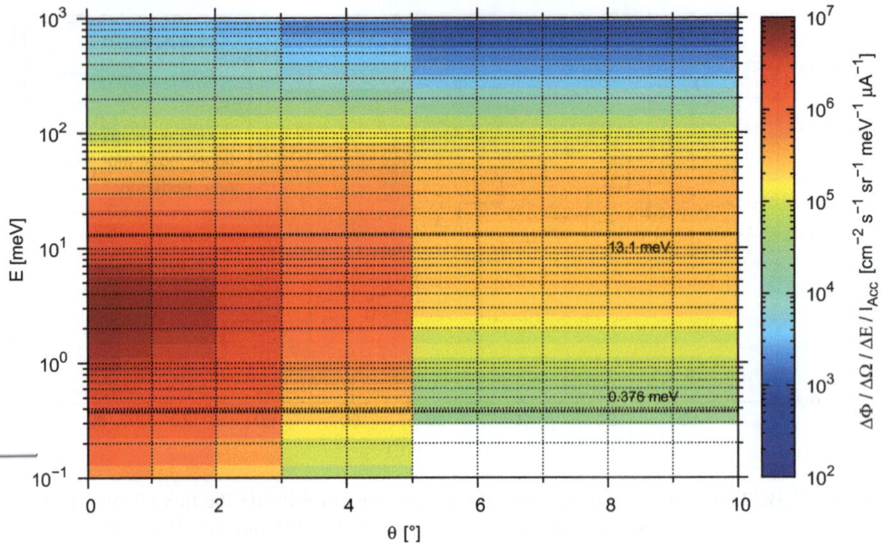

Fig. B.14 Energy-resolved brilliance of neutrons leaving the cold methane channel with $r = 3\,\text{cm}$, $\varphi = 10°$. The cold regime is marked by horizontal dotted lines

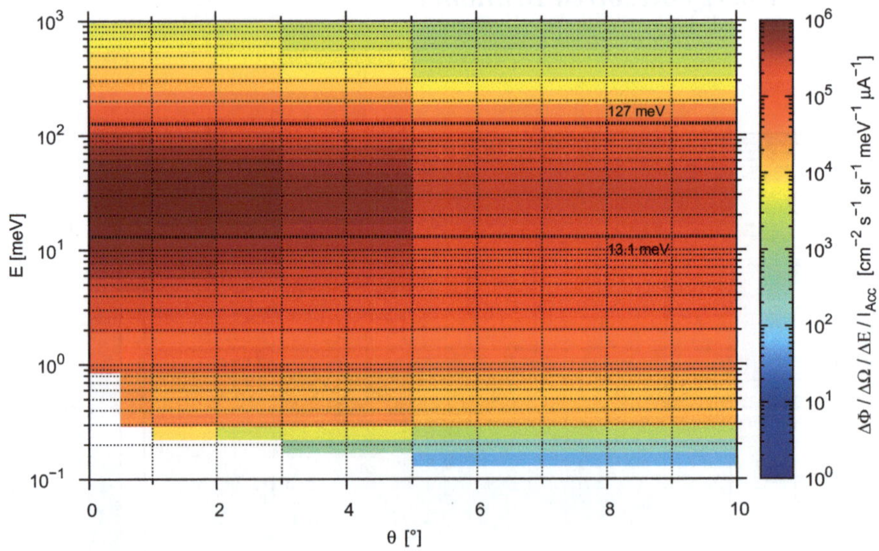

Fig. B.15 Energy-resolved brilliance of neutrons leaving the channel with $r = 3\,\text{cm}$, $\varphi = 90°$. The thermal regime is marked by horizontal dotted lines

B.2.4 Temporal Pulse Shapes

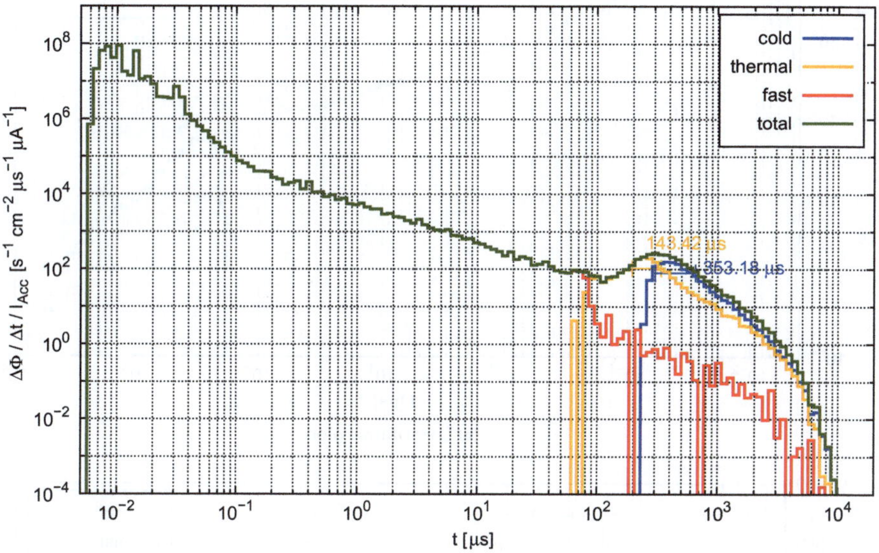

(a) Logarithmic plot.

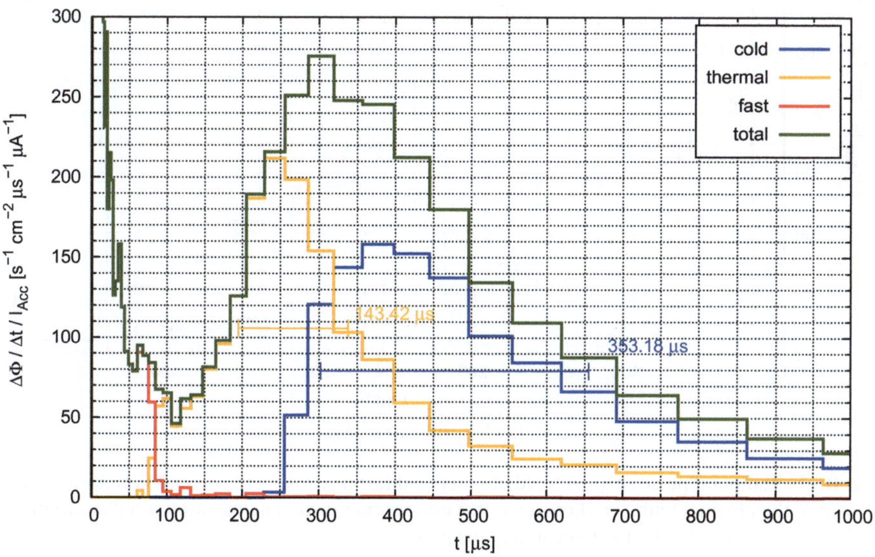

(b) Linear plot.

Fig. B.16 Temporal pulse shape of cold, thermal and fast neutrons with a maximum divergence of $\theta = 1°$ in the parahydrogen channel with $r = 1\,\text{cm}$, $\varphi = 45°$ considering instantaneous proton pulses according to Sect. 4.2.5.3. The FWHM is indicated in the plot

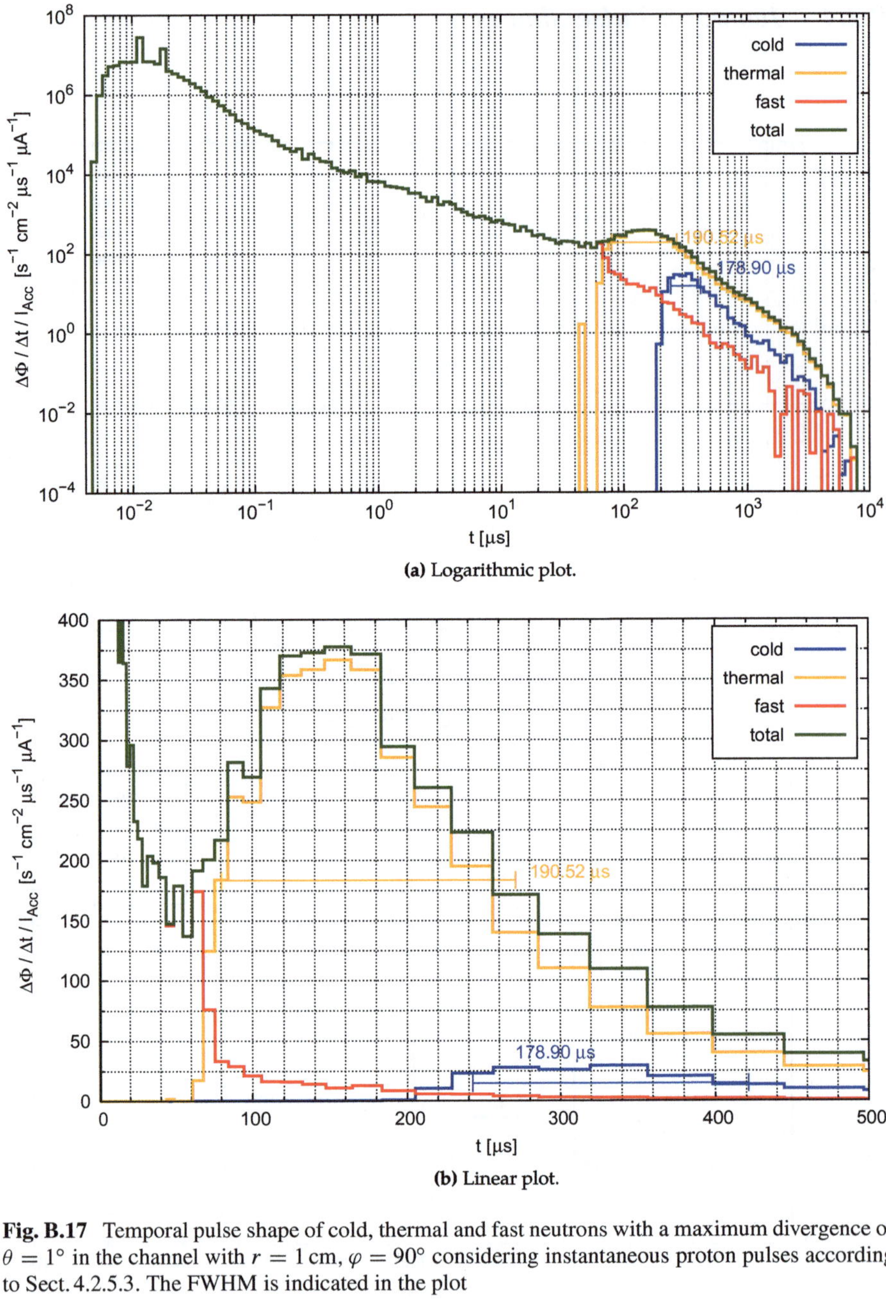

Fig. B.17 Temporal pulse shape of cold, thermal and fast neutrons with a maximum divergence of $\theta = 1°$ in the channel with $r = 1\,\mathrm{cm}$, $\varphi = 90°$ considering instantaneous proton pulses according to Sect. 4.2.5.3. The FWHM is indicated in the plot

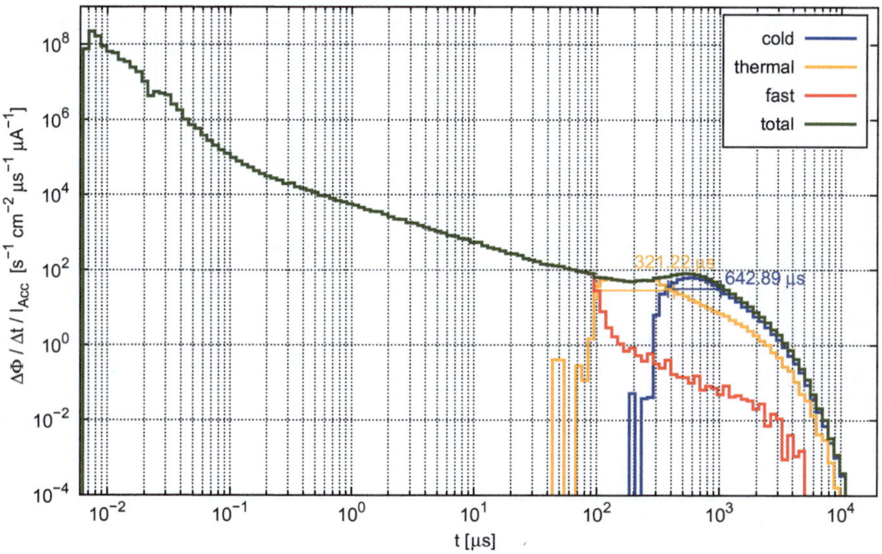

(a) Logarithmic plot.

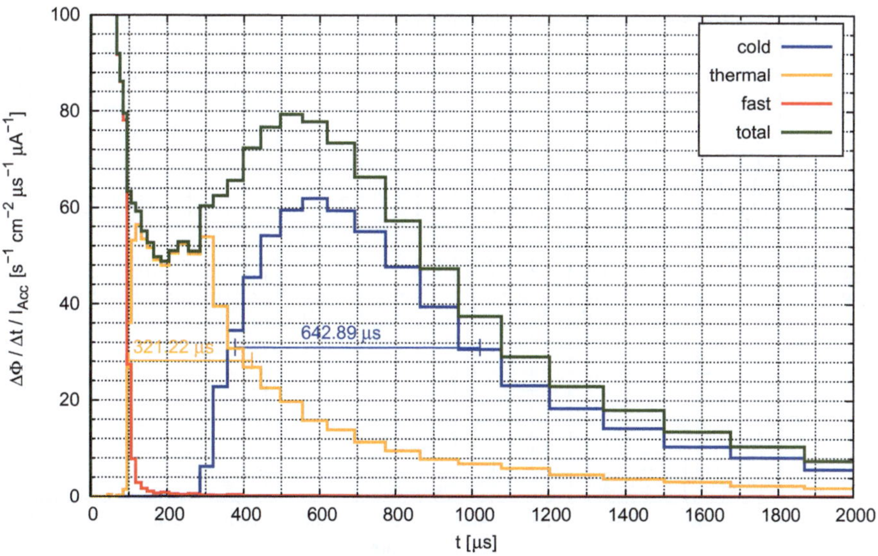

(b) Linear plot.

Fig. B.18 Temporal pulse shape of cold, thermal and fast neutrons with a maximum divergence of $\theta = 1°$ in the cold methane channel with $r = 3\,\text{cm}$, $\varphi = 10°$ considering instantaneous proton pulses according to Sect. 4.2.5.3. The FWHM is indicated in the plot

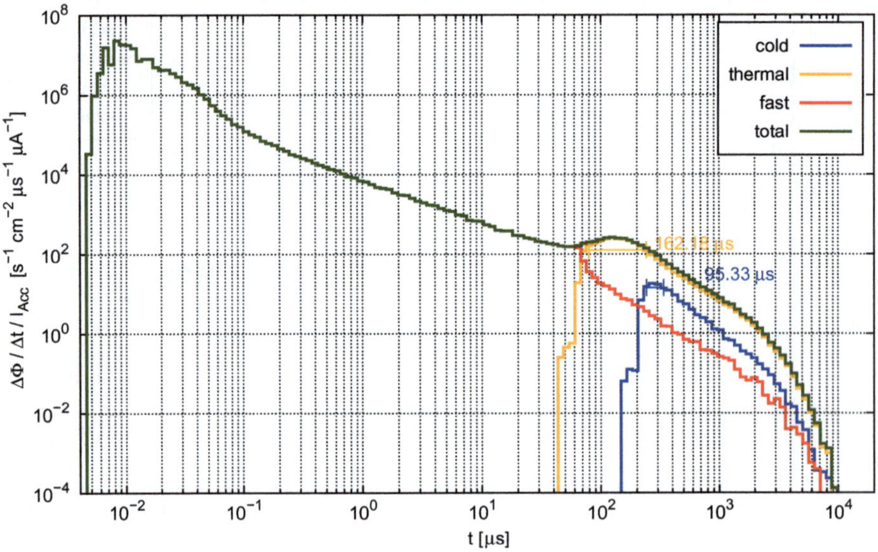

(a) Logarithmic plot.

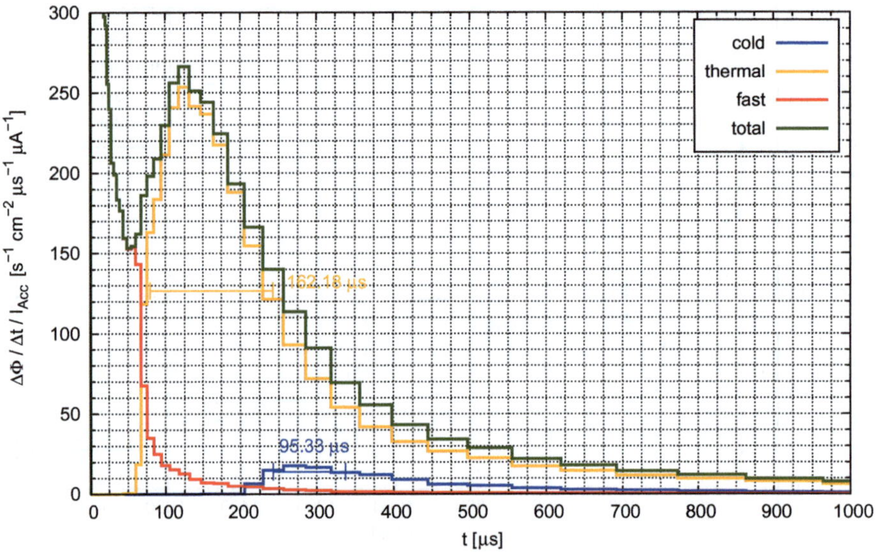

(b) Linear plot.

Fig. B.19 Temporal pulse shape of cold, thermal and fast neutrons with a maximum divergence of $\theta = 1°$ in the channel with $r = 3\,\text{cm}$, $\varphi = 90°$ considering instantaneous proton pulses according to Sect. 4.2.5.3. The FWHM is indicated in the plot

B.2.5 Brilliance for Different Reflector Thicknesses

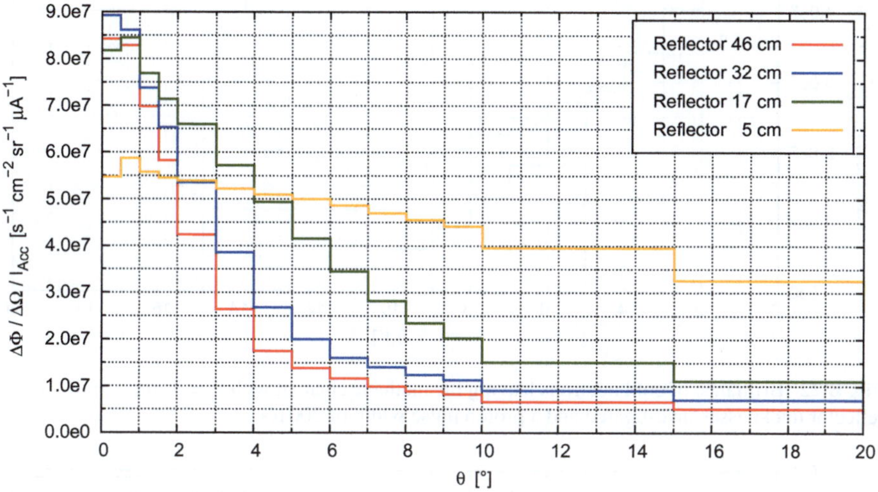

Fig. B.20 Brilliance of thermal neutrons leaving the channel with $r = 1$ cm, $\varphi = 90°$ for all reflector thicknesses investigated and a proton pulse width of 200 μs

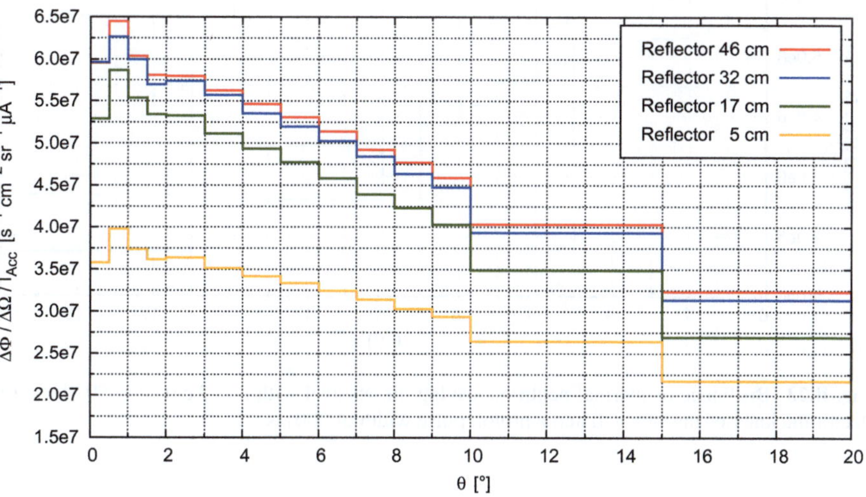

Fig. B.21 Brilliance of thermal neutrons leaving the channel with $r = 1$ cm, $\varphi = 120°$ for all reflector thicknesses investigated and a proton pulse width of 200 μs

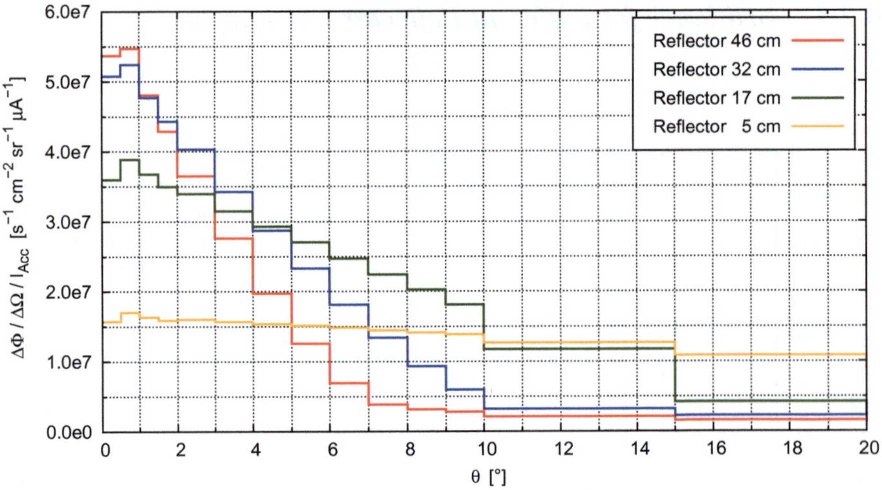

Fig. B.22 Brilliance of cold neutrons leaving the methane channel with $r = 3\,\text{cm}$, $\varphi = 10°$ for all reflector thicknesses investigated and a proton pulse width of 200 μs

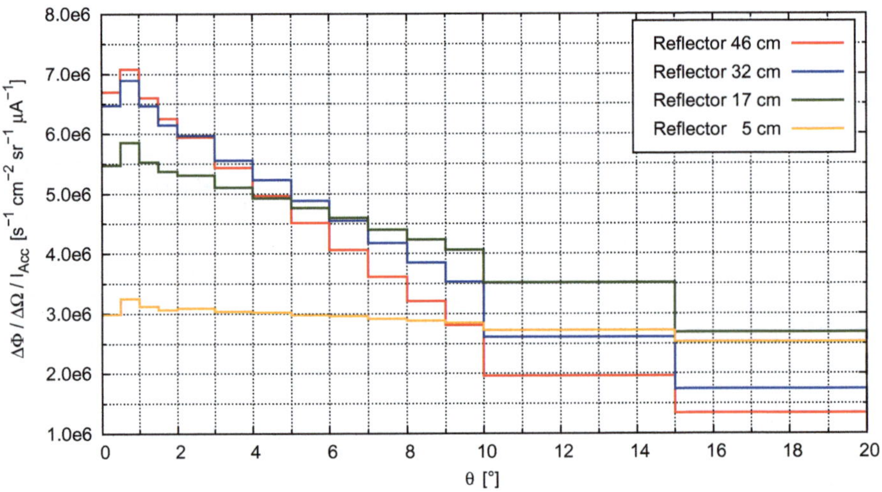

Fig. B.23 Brilliance of thermal neutrons leaving the channel with $r = 3\,\text{cm}$, $\varphi = 90°$ for all reflector thicknesses investigated and a proton pulse width of 200 μs

B.2.6 *Pulse Shapes for Different Reflector Thicknesses*

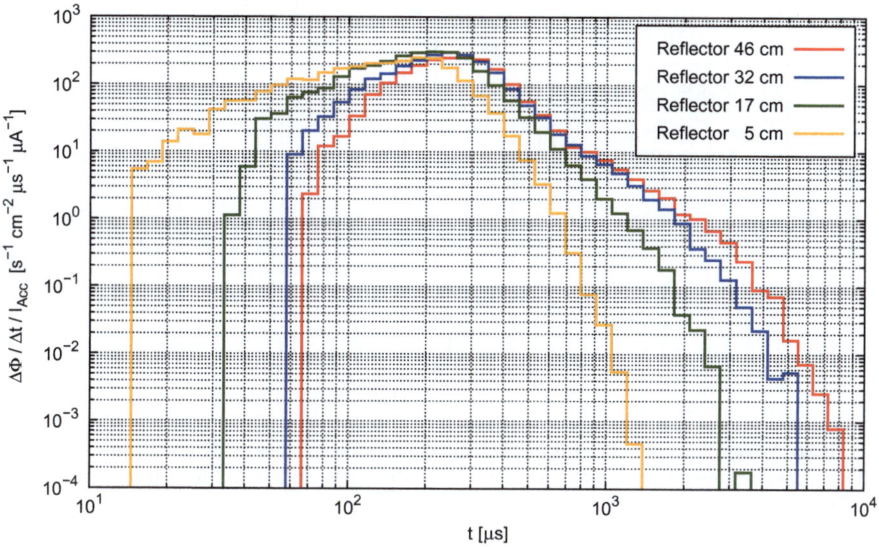

Fig. B.24 Pulse shape of thermal neutrons leaving the channel with $r = 1\,\text{cm}$, $\varphi = 90°$ for all reflector thicknesses investigated in a maximum divergence of $\theta = 1°$ for a proton pulse width of $200\,\mu\text{s}$

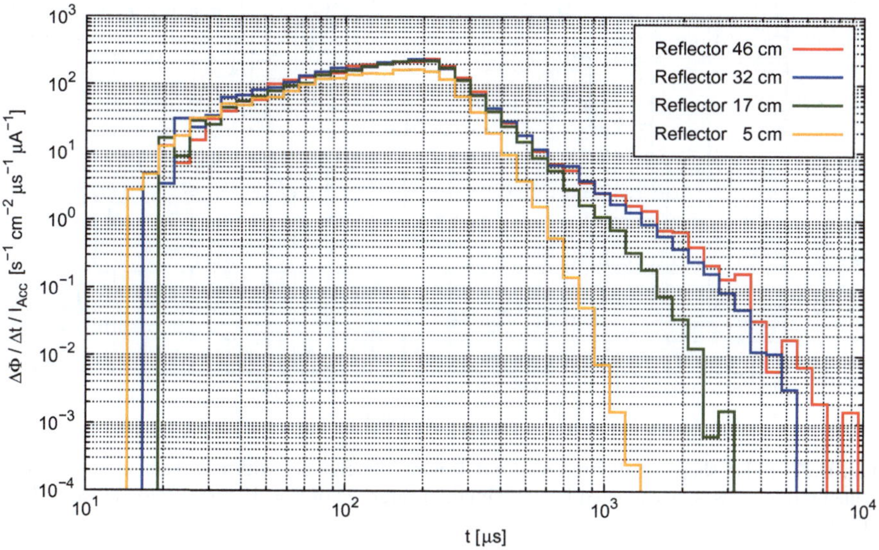

Fig. B.25 Pulse shape of thermal neutrons leaving the channel with $r = 1\,\text{cm}$, $\varphi = 120°$ for all reflector thicknesses investigated in a maximum divergence of $\theta = 1°$ for a proton pulse width of $200\,\mu\text{s}$

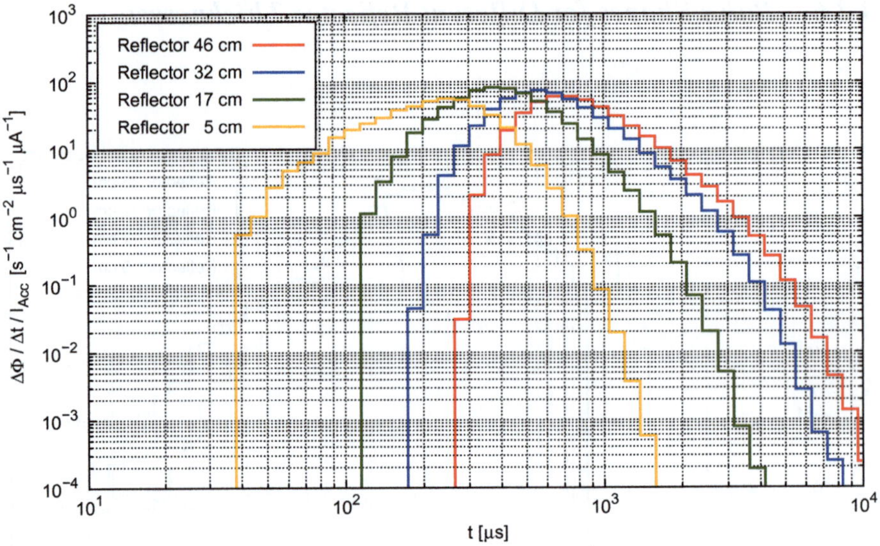

Fig. B.26 Pulse shape of cold neutrons leaving the methane channel with $r = 3\,\text{cm}$, $\varphi = 10°$ for all reflector thicknesses investigated in a maximum divergence of $\theta = 1°$ for a proton pulse width of 200 μs

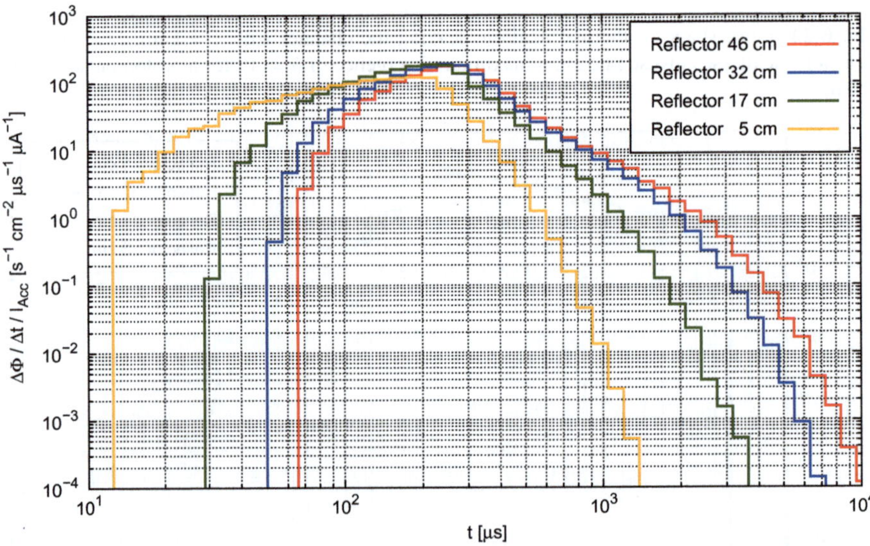

Fig. B.27 Pulse shape of thermal neutrons leaving the channel with $r = 3\,\text{cm}$, $\varphi = 90°$ for all reflector thicknesses investigated in a maximum divergence of $\theta = 1°$ for a proton pulse width of 200 μs

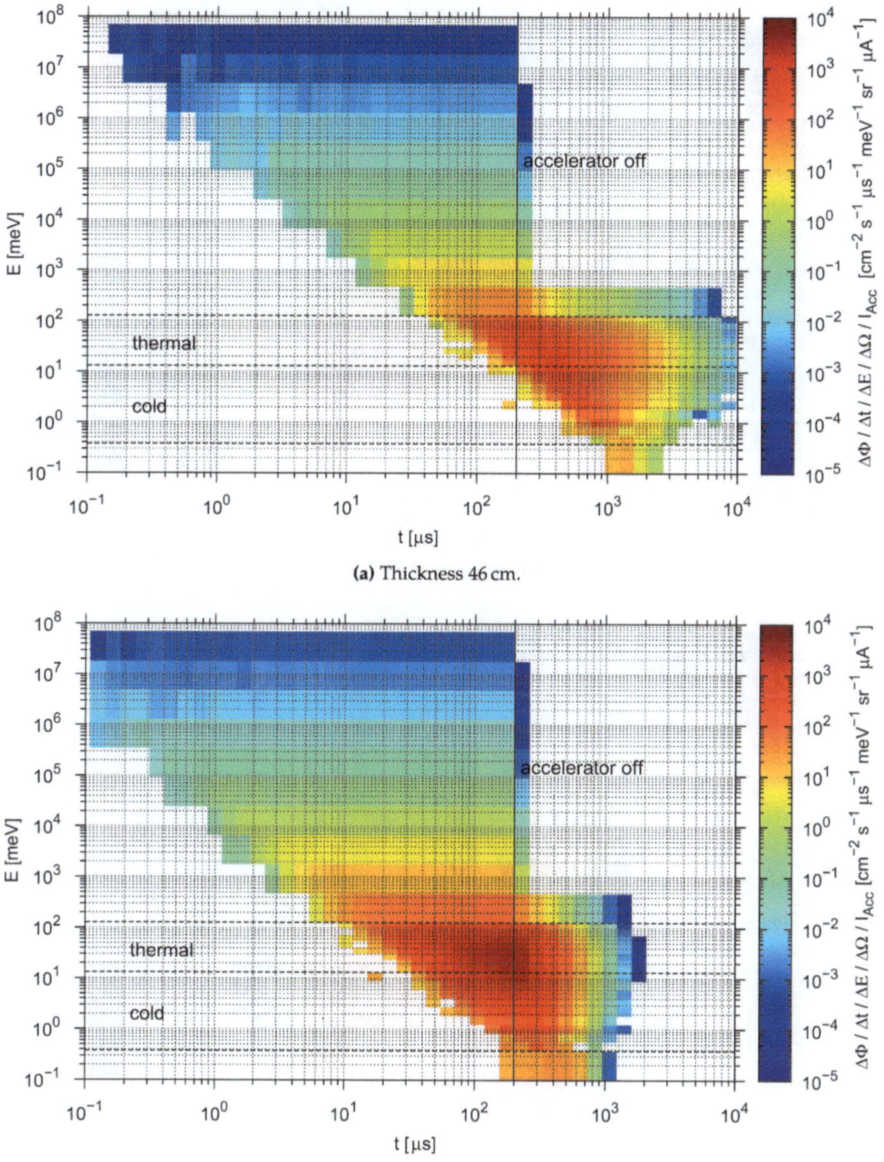

(a) Thickness 46 cm.

(b) Thickness 5 cm.

Fig. B.28 Energy-resolved pulse shape of the thermal flux channel with $r = 1$ cm, $\varphi = 90°$ for different reflector thicknesses and a proton pulse width of 200 μs. Neutrons with a divergence of $\theta \leq 5°$ are considered for better statistics

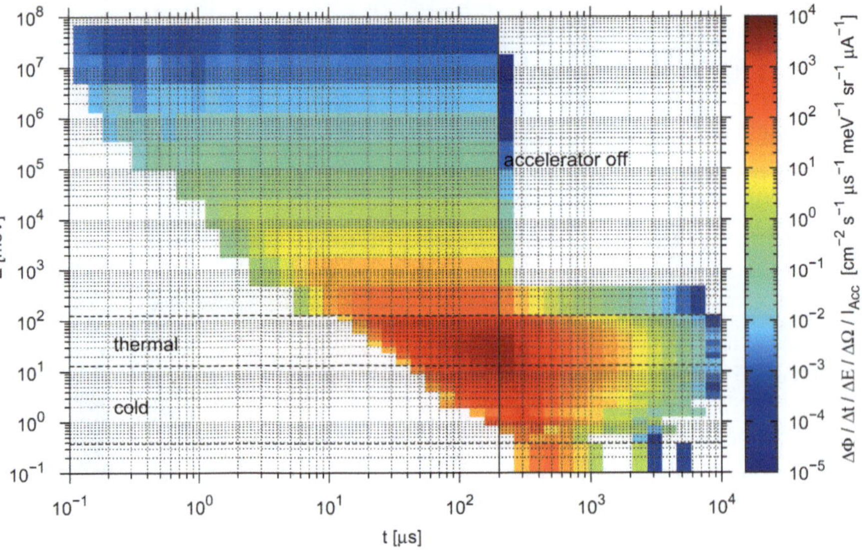

(a) Thickness 46 cm.

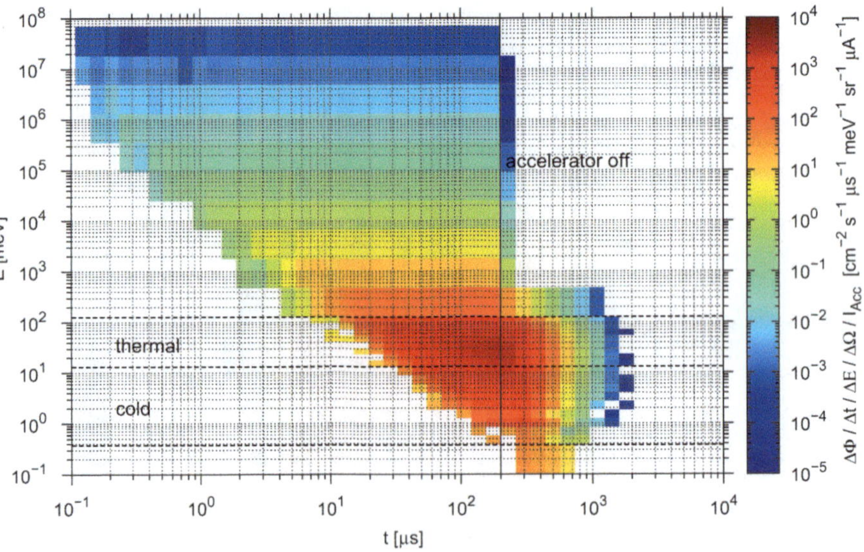

(b) Thickness 5 cm.

Fig. B.29 Energy-resolved pulse shape of the thermal flux channel with $r = 1$ cm, $\varphi = 120°$ for different reflector thicknesses and a proton pulse width of 200 μs. Neutrons with a divergence of $\theta \leq 5°$ are considered for better statistics

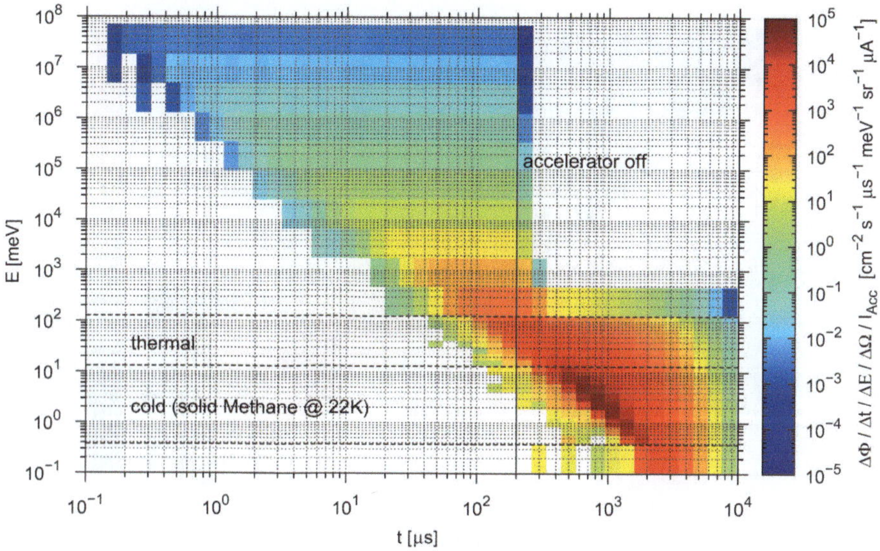

(a) Thickness 46 cm.

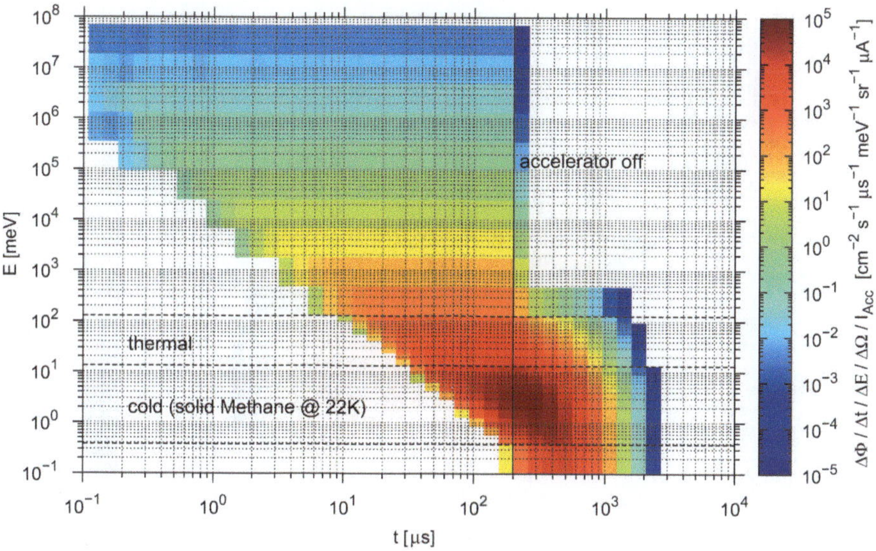

(b) Thickness 5 cm.

Fig. B.30 Energy-resolved pulse shape of the methane flux channel with $r = 3\,\text{cm}$, $\varphi = 10°$ for different reflector thicknesses and a proton pulse width of 200 μs. Neutrons with a divergence of $\theta \leq 5°$ are considered for better statistics

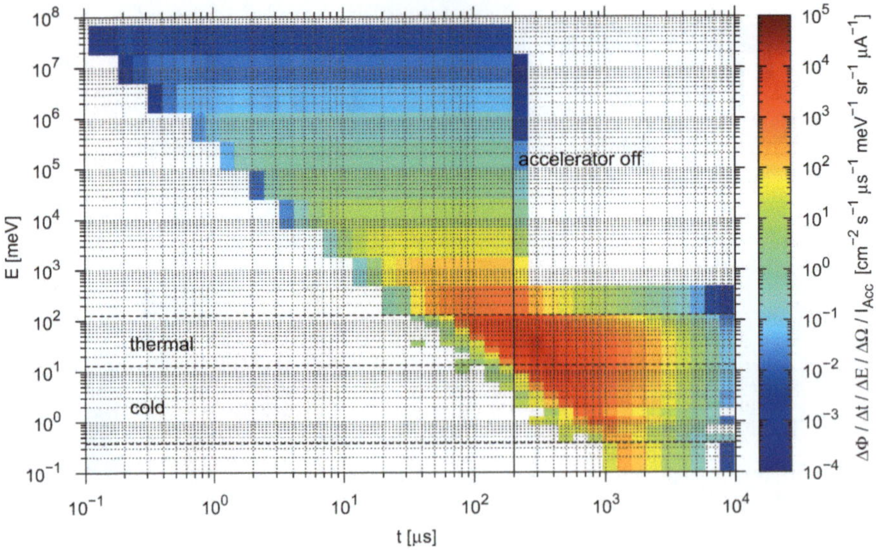

(a) Thickness 46 cm.

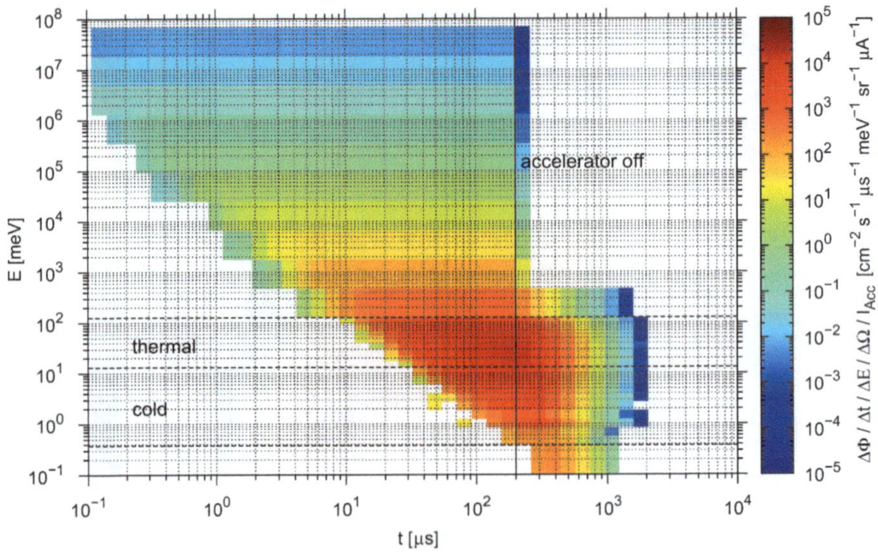

(b) Thickness 5 cm.

Fig. B.31 Energy-resolved pulse shape of the thermal flux channel with $r = 3$ cm, $\varphi = 90°$ for different reflector thicknesses and a proton pulse width of 200 μs. Neutrons with a divergence of $\theta \leq 5°$ are considered for better statistics

B.3 Prototype at the AKR-2 Reactor

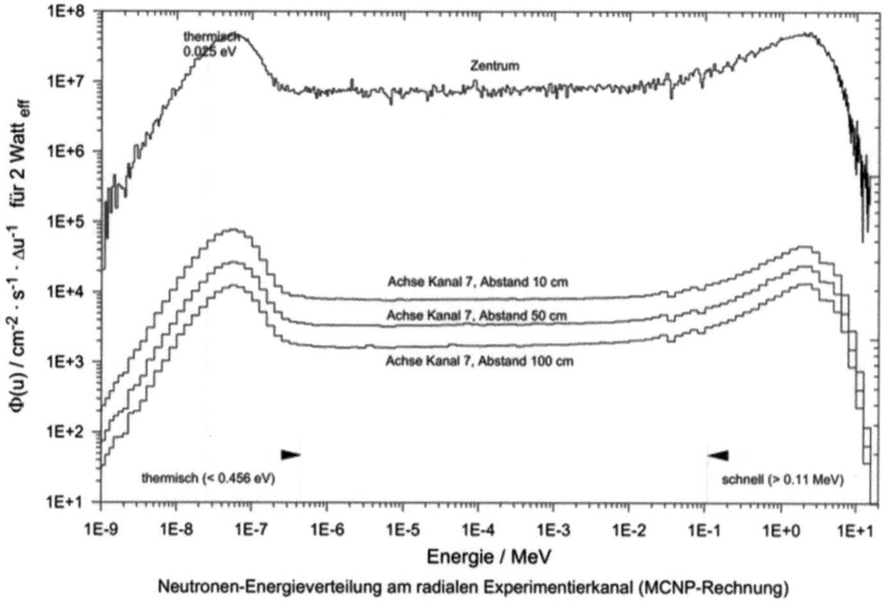

Fig. B.32 MCNP-computed and experimentally validated spectra in the reactor core and at different positions behind the exit of the radial channel on its axis. Calculations and measurements were conducted by [7]

Appendix C
Code and Scripts for Data Processing

C.1 PTRAC Source File

C.1.1 Generation of the Source File

If using the PTRAC command in the MCNP input an additional output file is created in which individual particle events are recorded. As only primary neutrons are of interest the recorded events must be filtered by using additional criteria [8]. The type of particles is restricted to neutrons by the TYPE=n keyword. The event type is set to "bank", see 2.5, via EVENT=bnk. To exclude secondary neutrons from (n, 2n) the lower energy cutoff for neutrons is set to 200 MeV so that all neutrons are terminated directly after creation. Additionally, the maximum number of collisions for neutrons which are to be recorded is set to 0. The MCNP input lines relevant for the correct generation of the PTRAC file read:

```
CUT:n j 200
PTRAC BUFFER=10000 FILE=ASC WRITE=ALL TYPE=n EVENT=bnk FILTER=0,NCP MAX
    =4E7
```

As multithreading and multiprocessing is not possible when using PTRAC, the MCNP calculations for generating the source file have to be performed in single jobs. To increase the calculation effectiveness, several serial jobs are started simultaneously, each using a different random number seed, see 2.5. In this way, different PTRAC outputs are generated, which can be easily concatenated. Bash scripts are used to start the individual jobs with different random number seeds, to process the PTRAC outputs and to generate the actual source file. These scripts are explained hereafter.

PTRAC Card to Record Particle Termination Events The PTRAC feature is also used to record termination events of the primary ions in the target, as discussed in Sects. 4.1.2.2 and 4.1.2.3. The corresponding PTRAC card for the MCNP input reads:

```
PTRAC BUFFER=1000 FILE=ASC WRITE=ALL TYPE=h EVENT=ter MAX=1E8
```


J. P. Dabruck, *Target Station Optimization for the High-Brilliance Neutron Source HBS*, Springer Theses, https://doi.org/10.1007/978-3-030-05639-1

Extract from PTRAC File An excerpt from the additional output encompassing the requested `PTRAC` data is listed hereafter. The first ten lines represent the header providing general information about the problem. Afterwards all histories complying to the specifications in the `PTRAC` card are listed, whereby the first line, e.g., line 11 of the first entry, of each entry contains the corresponding history number during the MCNP run and the event type `2030` which denotes a bank event where the banked particle is a neutrons produced by a light ion. The `9000` in the following line indicates that this event is the last event of the history in question. This line also contains additional information on the event like the material ZAID, the particle type, the cell number, the material number, and the number of collisions of the history. The actual particle data is stored in the next line as `XXX, YYY, ZZZ, UUU, VVV, WWW, ERG, WGT, TME`, cf. Table C.1. In history `140` two neutrons are produced by one initial proton.

```
1      -1
2   mcnp   6.mpi                    06/18/15 08/08/16 14:24:01
3   c
4      1.4000E+01 1.0000E+00 1.0000E+04 1.0000E+00 1.0000E+00 1.0000E+00 2.0000
           E+00 1.0000E+00 1.0000E+00 3.0000E+00
5      0.0000E+00 0.0000E+00 2.0000E+01 1.0000E+00 4.0000E+07 0.0000E+00 0.0000
           E+00 0.0000E+00 0.0000E+00 1.0000E+00
6      1.0000E+00 0.0000E+00 1.0000E+00 2.0000E+00 0.0000E+00 0.0000E+00 0.0000
           E+00 0.0000E+00 0.0000E+00 0.0000E+00
7        3    7    9    8    9    8    9    8    9    8    9    0    4    0    0    0
                 0    0    0    0
8       1   2   3   7   8   9  16  17  18  19  20  21  22  23  24  25  26  27  28   7
            8  10  11  16  17  18  19  20  21  22
9      23  24  25  26  27  28   7   8  12  13  16  17  18  19  20  21  22  23  24  25
           26  27  28   7   8  10  11  16  17  18
10     19  20  21  22  23  24  25  26  27  28   7   8  14  15  16  17  18  19  20  21
           22  23  24  25  26  27  28
11          25       2030        1
12        9000          3       4009        -1         1         1         1         0
13    -0.62683E-03 0.10440E-02 -0.20211E+01 -0.58814E+00 -0.63664E+00 0.49878E
          +00 0.78152E+00 0.10000E+01 0.53128E-01
14          42       2030        1
15        9000          3       4009        -1         1         1         1         0
16     0.17818E-03 0.85205E-03 -0.22417E+01 0.34453E+00 0.87363E+00 0.34362E+00
           0.85024E+01 0.10000E+01 0.50604E-01
17          46       2030        1
18        9000          3       4009        -1         1         1         1         0
19     0.10086E-01 -0.18248E-01 -0.14135E+01 -0.61464E+00 0.41464E-01 0.78772E
          +00 0.13309E+01 0.10000E+01 0.61102E-01
20          64       2030        1
21        9000          3       4009        -1         1         1         1         0
22    -0.11083E-01 0.34542E-01 -0.13867E+01 0.43731E+00 0.86541E+00 0.24459E+00
          0.17240E+02 0.10000E+01 0.61592E-01
23          93       2030        1
24        9000          3       4009        -1         1         1         1         0
25     0.00000E+00 0.00000E+00 -0.24109E+01 -0.47870E+00 0.44627E+00 0.75610E
          +00 0.37142E+02 0.10000E+01 0.48751E-01
26         126       2030        1
27        9000          3       4009        -1         1         1         1         0
28     0.90931E-02 0.18371E-01 -0.11247E+01 0.66577E+00 -0.52584E+00 0.52938E
          +00 0.20840E+01 0.10000E+01 0.66397E-01
29         140       2030        1
30        2030          3       4009        -1         1         1         1         0
31    -0.14329E+00 -0.12661E+00 -0.12591E+01 -0.19159E+00 -0.46182E+00 -0.86604
          E+00 0.15611E+01 0.95066E+00 0.64963E-01
32        9000          3       4009        -1         1         1         1         0
```

```
0.29365E-02 -0.16939E-01 -0.14845E+01 -0.29158E-01 -0.31713E+00 0.94793E+00 0.22298E+02 0.10000E+01 0.60061E-01
```

Extract from Source file The PTRAC file is parsed to extract the particle parameters line by line as shown in the excerpt from the generated source file. The first line is a header containing the particle type, the number of entries and lines, respectively, of the file as well as the number of the cell where the particles are to be started.

```
1 1996621 1
-6.26830e-04 1.04400e-03 -2.02110e+00 -5.8813981e-01 -6.3663980e-01 4.9877984e-01 7.81520e-01 5.31280e-02
1.78180e-04 8.52050e-04 -2.24170e+00 3.4452914e-01 8.7362781e-01 3.4361914e-01 8.50240e+00 5.06040e-02
1.00860e-02 -1.82480e-02 -1.41350e+00 -6.1463865e-01 4.1463909e-02 7.8771827e-01 1.33090e+00 6.11020e-02
-1.10830e-02 3.45420e-02 -1.38670e+00 4.3731027e-01 8.6541053e-01 2.4459015e-01 1.72400e+01 6.15920e-02
0.00000e+00 0.00000e+00 -2.41090e+00 -4.7870052e-01 4.4627049e-01 7.5610083e-01 3.71420e+01 4.87510e-02
9.09310e-03 1.83710e-02 -1.12470e+00 6.6576981e-01 -5.2583985e-01 5.2937985e-01 2.08400e+00 6.63970e-02
-1.43290e-01 -1.26610e-01 -1.25910e+00 -1.9158907e-01 -4.6181776e-01 -8.6603579e-01 1.56110e+00 6.49630e-02
2.93650e-03 -1.69390e-02 -1.48450e+00 -2.9158103e-02 -3.1713112e-01 9.4793336e-01 2.22980e+01 6.00610e-02
```

C.1.2 The Source Subroutine

If no source definition is specified by usage of the cards SDEF, KCODE, or SSR in the input file, MCNP calls the subroutine SOURCE which is empty by default. For custom source definitions the user has to implement this subroutine in which the parameters of the source particles are assigned.

Mandatory particle parameters are given in Table C.1. Only the parameters ERG, TME, UUU, VVV, WWW, XXX, YYY, ZZZ are read from the source file while IPT $\equiv$ 1 (1 for neutrons), WGT $\equiv$ 1, ICL $\equiv$ 1, and JSU $\equiv$ 0 are hard-coded in the subroutine. It has to be taken into account that MCNP performs a cross check of the source definition during which the coordinates of the starting particle are compared to the cell number ICL. If the source coordinates are not located within the cell ICL MCNP is terminated with a *bad trouble* error. For this reason, a change in the target geometry always requires the generation of a new source file.

The SOURCE subroutine is listed hereafter. In the lines 3–11 the needed modules are called and some variables are defined. A function getUnit to read a set of parameters is defined in lines 16–31. The subroutine init defined in lines 33–67 opens the external source file and allocates the parameters to an two-dimensional array by using the function getUnit. The actual source subroutine is located in lines 70–104. In this subroutine the particle parameters stored in the array are assigned to the MCNP-internal variables for the source particles as given in Table C.1. The source subroutine is MCNP-internally embedded in a *particle loop* which runs up to the number of nps. The variable counter defines which line of the

Table C.1 Source variables which have to be set in the `SOURCE` subroutine for each source particle

Source parameter	Description
`ERG`	Energy of the particle in MeV
`TME`	Time when the particle started in shakes
`UUU, VVV, WWW`	Flight direction of the particle
`XXX, YYY, ZZZ`	Position of the particle
`IPT`	Type of the particle
`WGT`	Statistical weight of the particle
`ICL`	Cell where the particle started
`JSU`	Surface where the particle started (0 if not started on a surface)

array is to be used for the current starting particle and is increased according to the MCNP-internal particle loop. If `counter>nps`, `counter` is set to zero and the first entry of the source file is used for the next history.

```
module source_data

  use mcnp_params, only:i4knd, dknd
  use mcnp_global
  implicit none
  save

  integer(i4knd) :: counter = 1
  integer(i4knd) :: num_particles, in_cell, par_type
  real(dknd), dimension(:,:), allocatable :: particle_data
  logical, private:: data_read = .FALSE.
  !!$omp threadprivate(counter)
  contains

  integer function getUnit()
    implicit none
    integer(i4knd) :: unit
    logical        :: isOpen

    integer, parameter :: MIN_UNIT_NUMBER = 30
    integer, parameter :: MAX_UNIT_NUMBER = 99

    do unit = MIN_UNIT_NUMBER, MAX_UNIT_NUMBER
      inquire(unit = unit, opened = isOpen)
      if (.not. isOpen) then
        getUnit = unit
        return
      end if
    end do
  end function getUnit

  subroutine init()

    implicit none
    integer :: namchg
    integer(i4knd) :: unt, stats, i

    if ( data_read == .FALSE. ) then
      !$omp single
      unt = getUnit()
```

```
      open(unit=unt, file='particle.input', iostat=stats, status='old',
           action='read')

      if(stats > 0) then
        call expirx(1,'source_data','could not open source particle
             information file')
      end if

      read(unt, *, iostat=stats) par_type, num_particles, in_cell
      ! convert problem name of cell tp "programm" name (array index)
      in_cell = namchg(1, in_cell)

      if(stats > 0) then
        call expirx(1,'source_data','read error')
      end if

      allocate(particle_data(1:8, 1:num_particles), stat=stats)

      do i=1, num_particles
        read(unt, *) particle_data( : , i)
      end do

      close(unt)
      data_read = .TRUE.
      !$omp end single
      !$omp barrier
    end if
  end subroutine init
end module source_data

subroutine source

  use mcnp_global
  use mcnp_debug
  use mcnp_random
  use source_data

  implicit none
  call init()

  IPT = par_type
  WGT = 1.0
  ICL = in_cell
  JSU = 0

  !$omp critical

  XXX = particle_data(1, counter)
  YYY = particle_data(2, counter)
  ZZZ = particle_data(3, counter)
  UUU = particle_data(4, counter)
  VVV = particle_data(5, counter)
  WWW = particle_data(6, counter)
  ERG = particle_data(7, counter)
  TME = particle_data(8, counter)

  counter = counter + 1

  if (counter > num_particles) then
    counter = 1
  end if

  !$omp end critical

end subroutine source
```

C.2 Data Binning Tool

The binning tool for analyzing the source file introduces two one-dimensional arrays which describes the bin boundaries defined in the file `grid.input` (lines 8–53) and a two-dimensional array whose elements correspond to the actual bin values. When passing through the source file line by line, the parameters to be binned are compared with the first two arrays and the corresponding element of the two-dimensional array is incremented by 1 for a match as achieved by lines 73–100. The output of the bin boundaries and the associated bin values occurs in the lines 101–110.

```
1  package binning;
2  import java.io.*;

4  public class Binning {

6      public static void main(String[] args) {

8          String grid = "grid.input";
9          String p = "particle.input";
10         String out = "binning.output";
11         try {
12             int sp1 = 0;
13             int sp2 = 0;
14             // int binMittelpunkt = 0, binBreiteX = 0, binBreiteY = 0;
15             double[] bin1 = null;
16             double[] bin2 = null;
17             long[][] binning;
18             String info1 = "Info1";
19             String info2 = "Info2";
20             FileInputStream gridStream = new FileInputStream(grid);
21             DataInputStream gridin = new DataInputStream(gridStream);
22             BufferedReader br1 = new BufferedReader(new InputStreamReader(
                       gridin));
23             String gridline1;
24             int k = 1;

26             while ((gridline1 = br1.readLine()) != null) {
27                 if (k == 1) {
28                     info1 = gridline1;
29                 } else if (k == 2) {
30                     gridline1 = gridline1.trim();
31                     gridline1 = gridline1.replaceAll("\\s+", " ");
32                     String[] line = gridline1.split(" ");
33                     sp1 = Integer.parseInt(line[0]);
34                     sp2 = Integer.parseInt(line[1]);
35                 } else if (k == 3) {
36                     gridline1 = gridline1.trim();
37                     gridline1 = gridline1.replaceAll("\\s+", " ");
38                     String[] line = gridline1.split(" ");
39                     bin1 = new double[line.length];
40                     for (int i = 0; i < line.length; i++) {
41                         bin1[i] = Double.parseDouble(line[i]);
42                     }
43                 } else if (k == 4) {
44                     gridline1 = gridline1.trim();
45                     gridline1 = gridline1.replaceAll("\\s+", " ");
46                     String[] line = gridline1.split(" ");
47                     bin2 = new double[line.length];
48                     for (int i = 0; i < line.length; i++) {
49                         bin2[i] = Double.parseDouble(line[i]);
50                     }
51                 }
52                 k++;
```

```
            }
            binning = new long[bin1.length][bin2.length];

            for (int i = 0; i < bin1.length; i++) {
                for (int j = 0; j < bin2.length; j++) {
                    binning[i][j] = 0;
                }
            }
            gridin.close();

            PrintWriter bout = new PrintWriter(out);
            FileInputStream pStream = new FileInputStream(p);
            DataInputStream pin = new DataInputStream(pStream);
            BufferedReader br2 = new BufferedReader(new InputStreamReader(
                 pin));
            String gridline2;

            double tmp1, tmp2;
            int b1;
            int b2;

            if ((gridline2 = br2.readLine()) != null) {
                info2 = gridline2;
                //System.out.println("Info 2: " + info2);
            }
            while ((gridline2 = br2.readLine()) != null) {
                b1 = 0;
                b2 = 0;
                gridline2 = gridline2.trim();
                gridline2 = gridline2.replaceAll("\\s+", " ");
                String[] line = gridline2.split(" ");
                tmp1 = Double.parseDouble(line[sp1 - 1]);
                tmp2 = Double.parseDouble(line[sp2 - 1]);
                for (int i = 0; i < bin1.length; i++) {
                    if (tmp1 > bin1[i]) {
                        b1++;
                    }
                }
                for (int i = 0; i < bin2.length; i++) {
                    if (tmp2 > bin2[i]) {
                        b2++;
                    }
                }
                if (b1 >= bin1.length | b2 >= bin2.length) {
                    continue;
                } else {
                    binning[b1][b2]++;
                }
            }
            bout.println("#" + info1);
            bout.println("#" + info2);
            for (int i = 1; i < bin1.length; i++) {
                for (int j = 1; j < bin2.length; j++) {
                    bout.println((bin1[i] + bin1[i - 1]) / 2 + " " + (bin2[j]
                          + bin2[j - 1]) / 2
                      + " " + binning[i][j] / ((bin1[i] - bin1[i - 1]) * (
                           bin2[j] - bin2[j - 1])));
                }
            }
            pin.close();
            bout.close();
        } catch (Exception e) {
            System.err.println("Error: " + e.getMessage());
        }
    }
}
```

References

1. F. Cantargi, J.R. Granada, Thermal neutron cross-section libraries for aromatic hydrocarbons, in *Nuclear Instruments and Methods in Physics Research Section B: Beam Interactions with Materials and Atoms 268.16*, pp. 2487–2491 (2010). ISSN: 0168-583X. https://doi.org/10.1016/j.nimb.2010.04.030. http://www.sciencedirect.com/science/article/pii/S0168583X10004209
2. J.R. Granada, F. Cantargi, J.I. Maquez Damian, Neutron cross sections libraries for methane in Phase II and solid deuterium, in J. Korean Phy. 59.2, Aug. 12, 2011, pp. 1076–1079. ISSN: 1976-8524. https://doi.org/10.3938/jkps.59.1076
3. R.E. MacFarlane. New Thermal Neutron Scattering Files for ENDF/B-VI Release 2. LA-12639-MS. Los Alamos National Laboratory, USA (1994). https://t2.lanl.gov/nis/publications/thermal.pdf (visited on 06/21/2017)
4. J.A. Young, J.U. Koppel, Slow neutron scattering by molecular hydrogen and deuterium. Phys. Rev. **135**(3A), A603–A611 (1964). https://doi.org/10.1103/PhysRev.135.A603
5. A. Nalbandyan, *Simulation of total neutron yields produced by proton beams on Be target. Report on Mini-Project. Institute of Nuclear Engineering and Technology Transfer* (RWTH Aachen University, May, 2016)
6. I. Tilquin et al. Experimental measurements of neutron fluxes produced by proton beams (23-80 MeV) on Be and Pb targets, in *Nuclear Instruments and Methods in Physics Research Section A: Accelerators, Spectrometers, Detectors and Associated Equipment 545*, pp. 339–343. 1–2 (2005) ISSN: 0168-9002. https://doi.org/10.1016/j.nima.2005.01.325. http://www.sciencedirect.com/science/article/pii/S016890020500522X
7. W. Hansen, R. Schneider, Erweiterung und Verifikation von TRAMO zur Lösung von gekoppelten Neutron/Gamma-transportproblemen und Überprüfung von Kerndatenbibliotheken. Final Report DFG Project. Project Number HA-3009/1-1. Technical University Dresden, Feb. 27, 2002
8. X-5 Monte Carlo Team. MCNP - A General N-Particle Transport Code, Version 5. LA-UR-03-1987. Tech. rep. Los Alamos National Laboratory. Feb. 1, 2008. https://laws.lanl.gov/vhosts/mcnp.lanl.gov/pdffiles/la-ur-03-1987.pdf
9. M.B. Chadwick et al. ENDF/B-VII.1 Nuclear data for science and technology: cross sections, covariances, fission product yields and decay data, in *Nuclear Data Sheets 112.12*, pp. 2887–2996 (2011). ISSSN: 0090-3752. https://doi.org/10.1016/j.nds.2011.11.002. http://www.sciencedirect.com/science/article/pii/S009037521100113X